Sarah Deborah Locke Stow

History of Mount Holyoke seminary, South Hadley, Mass.

During its first half century, 1837-1887

Sarah Deborah Locke Stow

History of Mount Holyoke seminary, South Hadley, Mass.
During its first half century, 1837-1887

ISBN/EAN: 9783337282103

Printed in Europe, USA, Canada, Australia, Japan

Cover: Foto ©berggeist007 / pixelio.de

More available books at **www.hansebooks.com**

A BIRD'S-EYE VIEW OF MOUNT HOLYOKE SEMINARY AND GROUNDS.
1887.

HISTORY

OF

MOUNT HOLYOKE SEMINARY,

SOUTH HADLEY, MASS.

DURING ITS

FIRST HALF CENTURY,

1837-1887.

BY

Mrs. SARAH D. (LOCKE) STOW,
OF THE CLASS OF 1859.

PUBLISHED BY THE SEMINARY,
1887.

TO

"The memory of Thy great goodness."

Ps. cxlv. 7.

PREFACE.

PROTESTANT education grounded in reverence for the Bible and faith in its teachings is an outgrowth of the Reformation. In the idea that "all men are created equal" it brought to New England civil and religious liberty, and a form of government which is safe and prosperous only in the hands of a people of intelligence and moral principle. That children and youth may be trained for good citizenship, there must be facilities for the education of women as well as colleges and higher schools for men. A sense of this need led to the founding of Mount Holyoke Seminary, and its origin may thus be traced back to the Reformation. Embodying the idea of a liberal and Christian education for women, on a permanent basis, it rose as a monument for the past, and a promise for the future. From this point of view the history of the two centuries before it was founded is yet to be written; for while we have modern histories, political, philosophical, and educational, we still lack one that duly takes into view the education of woman. Chapter I. of this volume may furnish toward it a few hints gathered chiefly from town histories and educational journals; but a full history can be written only after the things to be described in it shall have secured recognition.

Readers desiring fuller accounts of Miss Lyon and her colaborers are referred to the following works, from which material has been freely drawn: The Life and Labors of Mary Lyon; Hopkins, Bridgman & Co., Northampton: The Life of Mary Lyon; an abridgment of the former, with additions; American Tract Society, 150 Nassau street, New York: Recollections of Mary Lyon, by Fidelia Fiske; American Tract Society, Boston:

Daniel Safford; Congregational Publishing Society, Boston: Memorial Volume of Mount Holyoke Seminary, published in 1862: Life of Edward Norris Kirk, D. D.; Lockwood, Brooks & Co., Boston: Faith Working by Love,—Life of Fidelia Fiske; Congregational Sabbath-School and Publishing Society, Boston: The Use of a Life,—Memorials of Mrs. Z. P. Grant Banister; American Tract Society, 150 Nassau street, New York.

New matter has been added from Miss Lyon's correspondence and from other papers in the archives of the seminary.

As Mrs. Pease, secretary of the Memorandum Society, is collecting materials for a biographical record of Holyoke alumnæ, less space is given to individuals in these pages.

Chapters XIV., XV., XVI., and XVII. were written by Miss Mary O. Nutting, of the class of '52.

Grateful acknowledgments are due to Rev. J. M. Greene, D. D., for the use of his memorial sketch of the life of Rev. Roswell Hawks; to the hundreds of alumnæ who responded so cordially to the request for reminiscences or testimony; to present and former principals and teachers who have kindly aided in the work of preparation; to Rev. T. Laurie, D. D., and to Rev. Wm. S. Tyler, D. D., LL. D., for their valuable suggestions and careful revision; to the trustees, for their constant and cordial co-operation through their executive and library committees; and to all whose help has made it possible to issue the volume in season for the jubilee.

SOUTH HADLEY, April, 1887.

ERRATA.

On page 242, tenth line from bottom, "To the class of '71" should read "To the class of '68."

On page 349, fourth line from bottom, the date of death of Mrs. Mary C. (Whitman) Eddy is given as 1875; it should be 1874.

CONTENTS.

CHAPTER I.
EDUCATION FOR WOMAN IN THE UNITED STATES.
1636–1836.

First School in Boston—Why Colleges Were Founded—Harvard—Other Colleges before 1802—Law Requiring Universal Education—Penalty—Grammar and Primary Schools Required—Why Public Schools Were not for Girls—Decline in Education in the 18th Century—Effect of Revolutionary War—Admission of Girls to Public Schools—Law of 1789—Woman as Teacher—Wages—Proofs of Progress—Branches Taught—Academies—Girls' Schools—William Woodbridge—Mrs. Willard—Catharine Fiske—Catharine Beecher—Rev. Joseph Emerson—Miss Z. P. Grant—Expenses in Girls' Schools—Why More than in Colleges—Demand for Teachers—One Hundred Catholic Schools for Girls—One Hundred and Twenty Colleges—Mount Holyoke Seminary Incorporated, . 1

CHAPTER II.
THE LIFE OF MARY LYON.
1797–1849.

Birthplace—Ancestry—Inheritance—Education at Home—At School—Career as Teacher—Religious Character, ... 12

CHAPTER III.
MOUNT HOLYOKE SEMINARY ANTICIPATED.
1818–1834.

Mr. Emerson—Byfield Seminary—Miss Grant—Adams Academy—Buckland School—Ipswich Seminary—Efforts for Endowment—Effect of Failure—The New Plan—Its One Principle—Its Features, ... 27

CHAPTER IV.
PREPARATIONS.
1834–1836.

General Committee Appointed—First Thousand Dollars—Obstacles—Objections—Answers—Location—Name—Deacon Safford—Dr. Packard—Reliance upon Benevolence—Mr. Hawks—Circulars—Deacon Avery—Misses Maynard—Subscription Records, ... 39

CHAPTER V.
ERECTION OF FIRST BUILDING.
1836–1837.

Charter — Deacon Porter — Site — Corner-stone — Miss Grant—Effort in Boston—Wheaton Seminary—Prospectus—Progress in Building—Furnishing and Finishing, ... 61

CHAPTER VI.
THE OPENING.
1837–1838.

Contrast in Travel—Connecticut Valley—Hampshire County—South Hadley—Seminary Grounds—Seminary Building—Glance at Harvard—School Opens—Inconveniences—First Breakfast—Kindness of Villagers—Dedication, ... 78

CHAPTER VII.
THEORY PUT TO TEST.
1837–1838.

The Theory—Difficulties—Bread-making—Domestic Department—Plan of Organization—Success—Advantages—Misunderstood—Subsequent Modifications—Students of the First Year—Financial Success—First Anniversary—Diploma, ... 91

CHAPTER VIII.
METHODS ADOPTED.

Self-Help—Care for Health—Plan of Instruction—Use of Bible—Prayer—Specific Labor—Personal Influence—Teachers—Government—Cultivation of Conscience—Sense of Honor—Benevolence—Personal Responsibility—System Progressive—Caution. 103

CHAPTER IX.
METHODS ILLUSTRATED BY QUOTATIONS FROM PUPILS.

Miss Lyon's Personality—Pupils Free with Her—Her Hold on Them—Influence Abiding—Interest in Individuals—Watch over Mutual Influences—Enthusiasm and Thoroughness in Teaching—Notes of Criticism—Care for Health—Afternoon Exercises—Cure of Disorderly Habits—Dress—Sabbath Letter-writing—Counsels on Teaching—Motives—After a Death—Morning Addresses—Miss Lyon's Bible—Letter in Recapitulation—Told for a Memorial. 117

CONTENTS.

CHAPTER X.
GROWTH UNDER MISS LYON.
1837–1849.

Overflowing Numbers—Dr. Anderson's Address—Woman's Higher Education an Experiment—Plans for Enlarging the Building—Opposition—Success—Course of Study—Progress—Additional Advantages—French—Latin—Greek and Hebrew Desired, ... 138

CHAPTER XI.
GROWTH UNDER MISS LYON—CONTINUED.
RELIGIOUS HISTORY.
1837–1849.

Fast-days—First Revival—Blessings and Trials of 1840—Recess Meetings—Five Fruitful Years—Miss Lyon's Larger Desires—Praying for the World—Counsels upon Giving—Meeting at Norwich—Miss Fiske—Revival of 1843—"Missionary Offering"—New Interest in Missions—Other Teachers Leave—Revival of 1846—Mr. Condit's Death—Miss Rice Joins Miss Fiske—Miss Lyon's Last Vacation—Instructions—Illness—Death, ... 150

CHAPTER XII.
MISS CHAPIN'S ADMINISTRATION.
1849–1867.

Miss Whitman and Miss Hazen—Miss Chapin and Miss Spofford—Tribute of Dr. Fisher—Miss Jessup—Miss Tolman—Stewards—Improvements—North Wing—Observatory—Library Room—Death of Deacon Safford—Gymnasium—Extension of South Wing—Water Tower—Course of Study—Seminary in Wartime—Twenty-fifth Anniversary, ... 184

CHAPTER XIII.
MISS CHAPIN'S ADMINISTRATION—CONTINUED.
1849–1867.

Care for Souls—Dr. Kirk's First Visits—Miss Fiske's Return—Her Impressions of the Seminary—Missionary Reunion—Miss Beach—Revival of 1862–3—Call for Teachers—Miss Fritcher—Sabbath-school Party—Revival of 1863–4—Prominence of Prayer—Miss Pond—Death of Miss Fiske—Of Dr. and Mrs. Hitchcock—Miss Hopkins—Miss Chapin's Resignation—Mrs. Stoddard—Miss Norcross, ... 201

CHAPTER XIV.
ADMINISTRATION OF MISS FRENCH.
1867-1872.

The New Principal—Efforts to Introduce Steam-heating—How the Funds were Raised—Grant from the State—Donation of Mrs. Durant for Books—The Library Building—Longevity of Graduates from Various Institutions—Religious History of the Period—Teachers who Became Foreign Missionaries—Death of Mrs. Porter, of Hon. E. Southworth, and of Rev. Mr. Hawks—Resignation of Miss French and Miss Ellis, ... 221

CHAPTER XV.
ADMINISTRATION OF MISS WARD.
1872-1883.

Appointments Made—A Critical Period—Instruction in Modern Languages—Scientific Building Proposed—The School at Penikese—Obstacles—The New Colleges for Women—Continued Progress Indispensable—Tokens for Good—Dr. Kirk's Death—Mrs. Safford, Mrs. Eddy, and Mrs. Banister—Lectures on Art—Burning of the Church—Corner-stone of the New Building Laid—Anniversary in a Tent—Botanical Meeting—The Historical Sketch—Seminary Students at a Fire—Lectures of Joseph Cook, ... 235

CHAPTER XVI.
ADMINISTRATION OF MISS WARD—CONTINUED.
1872-1883.

Dedication of Williston Hall—Deacon Porter's Last Days—His Character—Advance in the Course of Study—The Paris Exposition—Memorandum Catalogue of 1878—Death of Mr. Sawyer—The Artesian Well—The Elevator—The Memorial Observatory—Transit of Venus Observed—The old Telescope in South Africa—Goodnow Park—Death of Colonel Rice—Mr. Kingman—Mr. Durant—Religious History of the Period—Seminary Teachers Laboring in Foreign Lands—Miss Clary—Resignation of Miss Ward—Resolutions Passed by the Trustees—Extract from the Paper of Dr. Tyler—Conclusion, . . 250

CHAPTER XVII.
ADMINISTRATION OF MISS BLANCHARD.
From 1883.

Events of the Year 1882-3—Miss Blanchard Elected Principal—Growth—Library Building Enlarged—Microscopical Outfit—Seminary Life at the Present Time—Religious History—

CONTENTS.

Teachers who have Gone to Foreign Work—Women Trustees—Change of Stewards—Financial Statements—Endowments Needed—Synopsis of the Present Course of Study, ... 263

CHAPTER XVIII.
STATISTICS.

Table of Attendance by Classes—Averages—Numbers from Different Places, ... 281

CHAPTER XIX.
RESULTS: TESTIMONY OF ALUMNÆ.

Christian Culture and Common Sense—One thing Holyoke does not do—It does Train for Usefulness—Influence of the Domestic Feature—Training Appreciated Gradually—Helps in the Work of Life—In Sorrow—Fortifies against Skepticism—Outlives Miss Lyon—Leads to Christ—Strengthens Christians—Cultivates Self-denial—Leaves its Stamp—Endears Private Prayer—Stimulates Interest in Missions—Alumnæ Send Daughters to Alma Mater—Gratitude of Loyal Hearts, . . . 288

CHAPTER XX.
ALUMNÆ AT WORK.

Memorandum Records—List of Foreign Missionaries—Alumnæ Associations, ... 318

CHAPTER XXI.
INSTITUTIONS MODELED AFTER MOUNT HOLYOKE SEMINARY.

In Persia—Tahlequah—Ohio—California—Michigan—Turkey—South Africa—Spain—Japan—Minnesota, ... 327

CHAPTER XXII.

Catalogue of Trustees—Teachers—Superintendents—Stewards—Anniversary Speakers—Pastors, ... 348

ILLUSTRATIONS.

STEEL PLATES.

	Page.
Seminary and Library,	1
Seminary Grounds, from Goodnow Park,	82
Portrait of Mary Lyon,	118
Lyman Williston Hall,	246

HELIOTYPES.

Bird's-eye View of Seminary and Grounds,	*Frontispiece.*
Birth-place of Mary Lyon,	20
The Parlors,	61
Art Gallery, interior,	103
Seminary Hall, interior,	132
Mineralogical Alcove,	147
Library, interior,	221
The Observatory,	235
Equatorial Telescope,	259
Botanical Room,	273
Zoölogical Laboratory,	300
Lecture room, Williston Hall,	315
Ornithological Alcove,	324

WOOD CUTS.

Birth-place Tablet,	12
Grave of Mary Lyon,	183

MOUNT HOLYOKE SEMINARY.
SOUTH HADLEY, MASS.
1887.

CHAPTER I.

EDUCATION FOR WOMAN IN THE UNITED STATES.

1636—1836.

MOUNT Holyoke Seminary was a pioneer in the higher education of woman. To understand the work of its founder, and to appreciate its own work, we must glance at the previous condition of education for woman in our country, and the influences which shaped the course of education in general.

The early settlers of New England, having left the old world in search of religious freedom, made it one of their first cares to educate their children for the sake of the well-being of both church and state.

How did they do this? At first, both in the Massachusetts and Plymouth colonies, instruction was given at home and by the parish ministers; but in 1635, within five years from the landing at the mouth of Charles River, provision had been made for the support of a teacher, a teacher selected, and a free school opened in Boston, then a small village of not more than twenty or thirty houses. Three years later, "dreading to leave an illiterate ministry to the churches when our ministers shall lie in the dust," Harvard College was founded, with the motto, *Christo et Ecclesiæ*. As Cotton Mather wrote years afterwards, "Our fathers saw that without a college to train an able and learned ministry, the church in New England must soon have come to nothing." Until Yale College was founded in 1700, Harvard furnished ministers and magistrates for all the New England colonies. The century of the Revolution was one of toil and difficulty, yet it witnessed the establishment of six other colleges by New England, and the rest of the Union was following her example. "Next to her political organization," says one writer, "we find each state looking out for a college as if it was the light of her eyes, or the right arm of her strength." By the second year of the present century there were twenty-eight; so soon and so far had the stars in our literary firmament outnumbered those upon our national flag.

"Lest there should be an illiterate ministry," colleges were founded. "Lest that old deluder, Satan, should keep men from the knowledge of the Scriptures," all children must be taught to read. Accordingly, by a law of Massachusetts passed in 1642, selectmen were to look after the children of those parents and masters who neglected to train them up in "learning and labor," and it was declared a "barbarism" not to teach them reading and a knowledge of the laws. Twenty shillings was the penalty for failure.

There were many schools in the colonies before the law requiring them in 1647. It was then ordered that every town with fifty families should provide a school where children should be taught to read and write; and that every town with one hundred families should provide a grammar school, the master thereof being able to instruct so far as to fit young men for college.

Thus within thirty years from the landing at Plymouth, was laid the foundation of our entire educational system, with its three grades of schools, essentially as they now exist.

Since the work of the grammar school was to prepare students for college, and the design of the college was to train men for the ministry, not many were found in either grade who had not that profession in view. Obviously colleges and grammar schools were not for girls. How was it with the primary grade? The law required the instruction of "all children," and the support of schools for "children." Girls were not mentioned. However it may have been at first, during most of the eighteenth century, town histories show that they did not ordinarily attend the public schools. There seems to have been no controversy on the subject. Their attendance was not thought necessary. At home, or in private schools kept by dames, they were taught to read and sew. It was deemed as important for them to read the Bible as it was for boys. The reading-book, in school or out, was the New England Primer. It contained the Shorter Catechism, which all children were required to commit to memory from beginning to end. Further learning than this girls were not supposed to need. Some learned to write; but when post-offices were few—and in 1790 there were but seventy-five in the country—correspondence was limited, and women in common life had little use for the pen. While everything worn in the family must be produced by the family, the education required by most was not to be gained at school, and it may be that girls were as well fitted for the part they were expected to fill in life, as boys for theirs. Their equal familiarity with the Bible and catechism prepared women to receive with as much profit as men, the intellectual and moral training afforded by Sabbath discourses which taxed the reasoning powers of all who heard.

The circumstances of the colonists gradually led to a general decline in education. For a generation or more before the revolutionary war, there is evidence that while all could read, there were people of respectability and influence throughout the country, who could not write. A large part of the wills left by men, some of whom had considerable property, and a larger part left by women, were signed with a cross. Early deeds in registrars' offices show that in many cases the wives of distinguished men were unable to write their own names.

The period of the Revolution was not favorable to improvement in this respect; and at its close there are said to have been ladies of high standing in Boston who could not read. But a new era in the education of woman soon dawned. Previously public schools with rare exceptions had been taught by "masters." In Northampton, before the war, farmers often taught English schools in the winter at from four to five dollars a month and boarded themselves at home.* After the war, young men found other occupations more lucrative, and women began to be employed in the summer schools. Girls began to attend. In many places it had been an unheard of thing for girls to be instructed by a master, and some towns were slow in allowing a change.

Boston did not permit them to attend the public schools till 1790, and then only during the summer months, when there were not boys enough to fill them. This lasted till 1822, when Boston became a city.† An aged resident of Hatfield, Massachusetts, used to tell of going to the school-house when she was a girl, and sitting on the doorstep to hear the boys recite their lessons. No girl could cross the threshold as a scholar. The girls of Northampton, Massachusetts, were not admitted to the public schools till 1792. In the Centennial *Hampshire Gazette* it was stated: "In 1788 the question was before the town and it was voted 'not to be at any expense for schooling girls.' The advocates of the measure were persistent, however, and appealed to the courts; the town was indicted and fined for this neglect. In 1792 it was voted by a large majority to admit girls between the ages of eight and fifteen to the schools from May 1 to October 31." They sometimes went to other towns to obtain the advantages of public schooling.

As late as 1828, a certain Otis Storrs was asked to take the town school in Bristol, Rhode Island, and allow girls to share his instructions. "Before this," the record adds, "girls did not go to the public schools."‡ The newer towns were apt to be in advance

---

* "History of Hadley," p. 426, note.
† "Quincy's Municipal History of Boston," p. 21.
‡ "History of Public Education in Rhode Island," Col. T. W. Higginson, p. 294.

of the older ones, and if unable to support additional schools for girls, often allowed them to attend with boys.

In 1789, Massachusetts passed a law indicating advance in several directions. Towns were authorized to establish school districts. Arithmetic, orthography, and the English language were to be taught in the common schools in addition to reading and writing. "Children in the most early stages of life" were to be admitted. One public school in a town, which was all that some towns had, would no longer suffice. In the need of more teachers, women were employed in schools where, besides morals, the requirements were only reading, "and writing if contracted for." Previously the law had recognized only masters as teachers. Hence only masters could collect wages. When women taught, their payment had been a voluntary matter. In the phrase "master or mistress," in the new law, is found the first legal recognition of woman as a teacher. It was not much money that she could collect. Twenty years later we read of teachers that received "one dollar a week and the privilege of working for board and earning another dollar." In Buckland, Massachusetts, 1799, with the wages of twenty-four weeks at fifty cents a week, a teacher bought for her wedding dress six yards of silk at two dollars a yard. In 1814, Mary Lyon began her career with seventy-five cents per week and board, and "boarded round,"—more briefly, "seventy-five cents a week and walked for board." This was twenty-five cents a week less than she had been receiving as her brother's housekeeper. In Connecticut this was a common price as late as 1830, according to the testimony of those who received it. It should be remembered, however, that in those summer schools only reading, writing, and morals were required to be taught. Much attention was given to "good manners." Many boys and all the girls brought work—straw-braiding, sewing, and knitting. Some had their "stints" for each half-day. Mothers and grandmothers are still living who show us samplers, laces, and other ornamental work done in school.

For years the summer schools in some parts of Massachusetts were supported by tuition fees, and not by tax upon the district; the principle of free schools for all, boys and girls, rich and poor, in its full application was but slowly adopted.

But the progress begun before the passage of the last named law, has never ceased. The number of studies was increased. At the opening of the present century we find geography introduced, and other reading-books taking the place of the New England Primer. In arithmetic each pupil took pride in ornamenting his manuscript book of rules and work with various styles of writing. Every one worked his own way and at his own rate, without recitation or examination. At first arithmetic and geography were taught only in the winter, for a knowledge of numbers, or ability to cast accounts, was deemed quite superfluous for girls. When Colburn's Mental Arithmetic was introduced, some of our mothers who desired to study it were told derisively, "If you expect to become widows and have to carry pork to market, it may be well enough to study mental arithmetic." But our mothers persevered and were not far behind their brothers in reaching the mathematical goal of the times—the Rule of Three. In spelling they were not at all behind; boys were quite as apt as girls to be "spelled down."

A growing interest in schooling for girls was manifest in another connection. In the decline of education in the eighteenth century, the lack of teachers for the grammar schools, and the desire for a higher order of instruction than common schools afforded in places where grammar schools were not required by law, led to the establishment of academies. They were for the benefit of all, but were not dependent like the grammar and common school upon local support and patronage. The first one was founded at South Byfield, Massachusetts, by bequest of William Dummer, who died in 1761.

Leicester Academy was incorporated in 1784; the one at Westford in 1793. These and others founded late in the century admitted girls, and thus, in addition to the greater advantages directly afforded, contributed their influence toward opening wider to them the doors of the common schools. Bradford Academy, when opened, in 1803, received both sexes. A separate department for girls was established in 1828. Eight years later the boys' department was closed and girls only have attended since. The first academy for girls only in New England was the Adams Academy, Derry, New Hampshire, incorporated June, 1823; the first in Massachusetts was Ipswich Academy, incorporated February, 1828. Abbot Academy, in Andover, was chartered in 1829.

Though these facts indicate the course of education in general, and the relative importance—or unimportance—attached to that of girls, there never was a time when the superior culture of the first colonists was not possessed by some of their descendants. Here and there were communities which constantly felt their influence, and these communities increased in number with the revival of interest in general education. There was no period when women of education and culture were not found in homes of superior intelligence and refinement.

The first school in New England, designed exclusively for the instruction of girls in branches not taught in the common schools, is said to have been an evening school conducted by William Woodbridge, who was a graduate of Yale in 1780. His theme on graduation was "Improvement in Female Education." Reducing his theory to practice, in addition to his daily occupation, he gave his evenings to the instruction of girls in Lowth's Grammar, Guthrie's Geography, and the art of composition. The popular sentiment deemed him visionary. Who, it was said, shall cook our food, or mend our clothes, if girls are to be taught philosophy and astronomy?

In Waterford, New York, in 1820, occurred the public examination of a young lady in geometry. It was the first instance of the kind in the state, and perhaps in the country, and called forth a storm of ridicule. Her teacher was Mrs. Emma Willard. Before her marriage in 1809, Mrs. Willard had been connected with a so-called female academy in Burlington, Vermont, and afterward in the same place, with a girls' boarding school, in which the higher branches were taught. "A Plan for Improving Woman's Education," from her pen, met the eye of Governor Clinton of New York, who persuaded her to remove to that state, and secured the passage of "An act to incorporate the proposed Institute at Waterford," and another, "To give female academies a share of the literary fund." This is believed to be the first law passed by any legislature, expressly for improvement in the education of woman. Instead of an institute at Waterford, the seminary at Troy was opened, over which Mrs. Willard presided from 1821 to 1838.

The year before Mrs. Willard left Troy, Miss Catharine Fiske, another worthy teacher, died in Keene, New Hampshire, where for twenty-three years young ladies from every state in the Union, more than twenty-five hundred in all, had received her instructions in Watts on the Mind, botany, chemistry, astronomy, and other studies.

It is said that every state in the Union sent representatives also to Miss Catharine Beecher's seminary in Hartford, Connecticut. This began in the chamber of a store, in 1822, with seven girls, none under twelve years of age, but soon numbered from one hundred to one hundred and sixty, and continued for ten years. Other teachers aided her, allowing a part of her time to be given to the invention of a system of calisthenics, and to the preparation of text-books for her pupils in arithmetic, mental and moral philosophy, and theology. Among other branches taught were the art of composition, history, Latin, and something of the art of teaching.

While these educators were doing a work in their respective states, the germs of a more permanent work were forming in Massachusetts. From 1818 to 1824, Rev. Joseph Emerson had been molding the intellectual and moral character of young women in his seminary at Byfield and Saugus. He had in all about one thousand pupils, and among them were teachers from different parts of New England, for the Normal School at Lexington, Massachusetts, the first one established in the United States, was not opened till July, 1839. One of these teachers was Miss Grant, who became the principal of Adams Academy, Derry, New Hampshire, and afterward of the seminary in Ipswich, Massachusetts. The connection of Mary Lyon with Mr. Emerson and Miss Grant will bring the work of each into fuller notice hereafter.

Many other young ladies' schools of reputation might be named, which flourished for a time but through lack of a financial basis inevitably declined on the death or departure of their founders. Without endowment their charges were necessarily high, and the expenses of girls at school were often double those of young men in college. According to Rev. Mark Hopkins, D. D., in an address in 1840, "in some cases the expense of sustaining a young lady in school for a year was more than double what was required to give a young man the advantages of a college course."

A course specially designed for ladies was provided at the Oberlin Collegiate Institute at its beginning in December, 1833. Mrs. Dascomb and Mrs. Henry Cowles, the first and second principals of that department, were pupils of Mr. Emerson and Miss Grant.

From the first, the importance to our country of having a class of well educated men had been so clearly seen, that state munificence and private liberality promptly endowed colleges and seminaries, securing the two results, permanence to the institution and moderate expenses to the student. But it was not thought important that there should be many well educated women, and accordingly girls were neither furnished by the state with facilities for separate study, nor even thought of in connection with colleges. Indeed during the first quarter of the present century, the common elements of education were generally considered sufficient for young women. The first symptom of progress had not been encouraging. It was an exaggerated estimate of the value of mere learning. Ornamental branches and so-called accomplishments "finished" the education of the comparatively few whose means allowed. "I spent a thousand dollars," said a father, "to educate my daughter. I would give another thousand to undo it all. She has been made vain, frivolous, and discontented with the plain, simple habits of home." It is not strange the mistaken impression prevailed that learning, instead of fitting young women for their appropriate duties, had a tendency to make them less domestic, less healthy, and less useful. But influences were already at work to bring about a change. Girls were as freely and as well taught as boys in the common schools, and in academies where they existed. Woman, legally recognized as teacher, had been constantly gaining in opportunity and ability. Of the teachers in the common schools of Massachusetts for 1837, the ratio of women to men was as five to three. The demand for schools and teachers was steadily increasing, not only in New England but in western states. Every successful school added to the previous demand. There were a few in our rapidly growing country whose eyes were open to the need. They saw the hope of the nation in the training of the young, and this training passing more and more into the hands of woman. To the ever constant moral molding by the mother was being added the mental training by both mother and teacher. And the Pope of Rome was looking on, not idly; Europe was sending millions of money across the sea to support schools for girls who were Catholics, or to become Catholics. In the summer of 1836, New York City received one hundred thousand dollars, for this object. There were then in the United States one hundred Catholic schools for girls. Rome knew that "the hand which rocks the cradle rules the world," and sought the control of that hand in America. A few Protestant observers, equally wise, sought to place that hand under the control of Christian intelligence. They desired permanent facilities for the liberal education of young women to be trained "for Christ and the church," and for this they toiled and prayed.

To realize the need was much. To supply that need was more. There was prejudice to be removed, indifference to be overcome, philanthropy to be roused, benevolence to be called into action. A leader was needed of a clear and well balanced mind, strong, inventive, earnest, and enthusiastic; of a spirit undaunted by opposition and prejudice, by obstacles known and unknown, by failures apparent or real; with a faith that could remove mountains of difficulty and trust in the dark; with a benevolent heart, moved by pure impulses, warm and large enough to make friends of opposers; and with a body equal to the demands of such a mind and heart.

When the time had come for such a work, such a leader appeared and the difficult work began.

In 1636, "The General Court of Massachusetts agreed to give one hundred pounds towards a college." Two hundred years passed, and there were in the United States one hundred and twenty colleges for young men, when in 1836, the legislature of Massachusetts granted a charter to Mount Holyoke Seminary, for young women.

CHAPTER II.

THE LIFE OF MARY LYON.

1797—1849.

IN the Alpine regions of Massachusetts, quite away from the usual lines of travel and yet of easy access to those who love the hills, lies the deserted site of a once busy home. You may approach it by a "wild winding way" through a maple grove, by a drive of several miles from the railway, or by a climb over the steep hill at the west, from whose summit, one thousand one hundred feet above the sea, the prospect stretches away over a billowy sea of mountains in different directions, to Greylock, Monadnock, and Wachusett. By whatever path the place is reached, one may apply to it the words written long ago by one familiar with the landscape, "A wild, romantic farm, made more to feast the soul than to feed the body."*— That it was not a barren spot, the same writer shows in her pictures of the home surrounded " by roses, pinks, and peonies that keep time with Old Hundred "; and of the garden where nothing ever died; its richly laden fruit trees; the sugar maples shading the musical brook; wild strawberries growing in richness and profusion; garden, grove, and field, yielding such generous supplies that there was always enough and to spare, even when young life most abounded in that mountain home.

Young and old have long since departed, and even the home is gone. Ferns are growing from the cellar wall. Red roses hide in the grass, and a few venerable fruit trees linger. But when these, too, are gone, the little brook will still sing on at the foot of the sentinel hills.

This deserted site in Buckland, four miles from the village by the road, but only a mile and a half by footpath over the hill, is the place where Mary Lyon was born, February 28, 1797. The trustees of Mount Holyoke Seminary have marked the spot by a bronze tablet inscribed with her name, and inserted in the rocky ledge. Mary was the sixth of eight children given to Aaron and Jemima (Shepard) Lyon. Her father, who died before she was seven years of age, was a man greatly beloved by all, and often sent for to pray with the sick and dying. Her mother's ancestry goes back in one line to Rev. Henry Smith, who came from England as early as 1636, and in another to Lieut. Samuel Smith, who sailed for New England in the *Elizabeth*, of Ipswich, in 1634. From Charlestown and Watertown, Massachusetts, they went to Wethersfield, Connecticut, where one was minister till his death in 1648, and the other for twenty years filled important offices in church and state. Differences arose, due in part to lax views respecting baptism and church membership, and all but

---

* "The Missionary Offering," p. 57.

six members of the church, "for the sake of peace and harmony," voted to remove from Wethersfield. In 1659 or 1660, about thirty left in a body with Rev. John Russell, their minister, and joined by a minority from the church in Hartford, holding similar views, were the first settlers of Hadley, then called Norwottuck. Among these were Lieut. Smith, his son Chileab, and some of the children of Rev. Henry Smith; Preserved, the grandson of the minister, married Mary, the granddaughter of the lieutenant, thus uniting the two families. Their son Chileab, bearing the name of his maternal grandfather, left before 1731, to become one of the first settlers of South Hadley. Chileab's Hill, a mile north of the village, still bears his name. An old well, near the present residence of Dennison Smith, belonged to the homestead where eleven of his twelve children were born. Jemima, the fifth, was the grandmother of Mary Lyon. Chileab Smith was a man of sterling qualities. The questions on which the churches of Hartford and Wethersfield were divided in the preceding century, led to much discussion in the region of Hadley. In this controversy Jonathan Edwards earnestly opposed the loose theory of church membership, and when leaving Northampton, said that no church and only one minister in Hampshire county was fully of his mind. Hampshire then comprised also the present counties of Hampden and Franklin. It is stated that the great-grandfather of Mary Lyon withdrew from the church in South Hadley, having had some difficulty with it "because he agreed with Mr. Edwards." In 1751 with two other families he began the settlement of Huntstown, now Ashfield. A pioneer like his fathers, he was like them foremost also in Christian work. In this place he began religious meetings at once and out of them grew a Baptist church. Its first pastor was his son Ebenezer, who preached from the age of nineteen to eighty-nine. During forty-five of these years he ministered at Ashfield, and was succeeded by his brother Enos, who was pastor forty years. Besides these two sons, ten ministers, and five daughters whose husbands were ministers, were counted in 1851 among his descendants. At the age of eighty he was himself ordained to the ministry. He lived to be ninety-two. His wife died at eighty-seven. The grandfather, whose name he bore, reached the age of ninety-five, his grandmother eighty-eight, his son of the same name one hundred years and seven months. Two other sons lived to be more than ninety years of age. His daughter Jemima on one occasion wrote that it had been seventy years since she "'listed a soldier for Jesus." Her husband was Deacon Isaac Shepard, of Ashfield, a man of eminent piety like his father before him. Their daughter Jemima Shepard became the wife of Aaron Lyon, September 2, 1784.

From such a line of long-lived ancestors,—from self-reliant, hard-working parents, in Christian love tenderly devoted to each other and to their children,—from her mother, a cheerful, capable woman, noted for her faith and prayers, and an angel of goodness among her neighbors,—Mary Lyon inherited a sound mind in a sound body, a buoyant temperament, a great, warm, trusting heart, with intense energy of body, mind, and soul. Nurtured in a home of rural simplicity and Christian sincerity, unfettered by custom and fashion, inhaling strength with the free mountain air, and gathering stores of wisdom from her mother's Bible, this blue-eyed girl, with fair skin, rosy cheeks, broad, high forehead, and masses of curling auburn hair, was laying up invaluable resources for after years. Skilled in all the household arts of that day, doing at fifteen a hired girl's work for a hired girl's wages while keeping house for her unmarried brother, and giving to the work head and heart as well as hands, she was educating herself for more important labors. Spinning now flax, now wool; weaving, netting, and embroidery gave variety to her work and increased her power of self-help. With two blue-and-white coverlets spun, dyed, and woven by her own hands, she afterward paid for a winter's board while in Ashfield Academy; and the blue fulled-cloth habit she wore in Ipswich and Derry was the product of her own spinning wheel and loom.

In a home where children were thrown on their own resources, and in a community where school privileges were few and genius was lightly valued, her eager intellect was safe from undue stimulus, and physical vigor was allowed free development. Exuberant spirits, a keen sense of the ludicrous, and a power of humorous description combined with overflowing kindness, made her society attractive. And while she greatly outstripped her schoolmates in their studies, they regarded her with admiration rather than envy. "Even then," says one of them, "she was so full of benevolence that we were all drawn to her."

Her school advantages were limited. When she was seven years old the district school which had been near her home was removed to a distance. Her occasional attendance afterward was doubtless supplemented by home studies. Sometimes she lived with families in Buckland or relatives in Ashfield where she could work for her board and be nearer school. At the age of twenty she first went to Sanderson Academy in Ashfield. Two quarters there in 1817 and one at Amherst Academy in 1818 alternated with terms of teaching near her home. In 1820 she was again in Sanderson Academy, of which she afterward said, "Here I was principally educated, here my mental energies were first awakened, and to this school I feel in no small degree indebted." The indebtedness perhaps consisted in part in receiving free tuition after the first term, although it has been said that "she heard recitations enough to more than balance her tuition."

Here, contrary to her principles in after life, she gave herself only four hours of sleep in twenty-four, counting study time too precious to be taken for sleep. Catalogues, each a single broad-sheet, found carefully preserved among her papers, suggest how she cherished the memory of those days.

While attending district school, in addition to the reading, writing, spelling, and needlework with which girls were usually expected to be satisfied, she presented the extraordinary request to be allowed to study grammar. The principles of the English grammar were so well stored in memory and that within four days, that when a Latin grammar was afterward given her in Sanderson Academy to keep her within reciting distance of her classes in other studies, she was but three days in mastering the book. Schoolmates forgot to study that afternoon, listening to the unfailing promptness of that long recitation, which is said to have lasted till after dark. Under the same teacher, Elijah H. Burritt, she calculated eclipses and made an almanac.

Between terms she was at one time in the family of Rev. Edward Hitchcock, then pastor in Conway, with whom she studied natural science, taking also lessons in drawing and painting from Mrs. Hitchcock. Her fondness for natural science was further gratified in Amherst Academy, where she specially delighted in chemistry, as she did later in lectures by Prof. Eaton on chemistry and natural philosophy, in Amherst and in Troy, New York. In these sciences she became an enthusiastic teacher, performing her own experiments. The same eager thirst for knowledge which these four or five terms of instruction had only increased, led her in 1821 to go to Byfield, a distance of three days' journey eastward,—for it was six years before the survey of the pioneer railway in Massachusetts. At first her plan was opposed. She was twenty-four years old; she had learning enough to teach; why should she go to Byfield? But her mother said, "Go." During her two terms there under the influence of Rev. Joseph Emerson, her views of education underwent a radical change. She had been seeking knowledge for the joy of acquiring it. At Byfield she learned to desire it as a means of usefulness. She had been cultivating the intellect mainly, but was now led to see that the heart also should receive due attention.

Years afterwards Mr. Emerson said that for mental power he considered Mary Lyon superior to any other pupil he had ever had in his seminary. With the consecration of this power, not only was the development of mind and heart more symmetrical, but she also illustrated a later saying of her own, "The same talent rises higher with consecration." In reviewing her life near its close she said that she owed more to Mr. Emerson than to any other instructor.

Her career as a teacher began when she was seventeen. From a list in her hand-writing of terms taught and attended, it appears that in the seven years before going to Byfield she had taught eleven terms in district schools in or near Buckland, varied with an occasional family or select school. Many times during her first term—of twenty weeks—she resolved if once safely through, never to attempt the work again. Others said, "She will never equal her sister Electa as a teacher." But each term so improved upon the preceding that her services were soon in eager request. She was the first to introduce into Buckland the study of geography with maps. After her return from Byfield, inspired with new motives and new views of education, she held for three years the place of assistant principal in the academy in Ashfield, a position which no woman had occupied before. Results fully justified the innovation. The next ten years she was associated with Miss Grant in Derry, New Hampshire, and Ipswich, Massachusetts, returning six winters for a school of her own in Ashfield or Buckland. In 1834 she left the schoolroom to devote herself to the enterprise for which twenty years of teaching, and indeed all her life, had been preparing her. Three years she toiled upon the problem given her to solve, and began to test its solution when she opened Mount Holyoke Seminary, November 8, 1837. And if "the honor of discovery belongs to him who first puts the novel idea into such working form as to make it of practical value," that honor surely belonged to Mary Lyon. In the words of President Hitchcock, "Until she made the effort it was not thought that founding schools for women was of consequence enough to be counted a benevolent enterprise, but she so convinced the church of its importance as to draw forth its pecuniary offerings in a greater degree than selfish motives had ever accomplished."

In religious character, Mary Lyon was by birthright a woman of faith, and her life has been called an added verse to the eleventh chapter of Hebrews. Born of believing parents, themselves rich in a heritage of faith, she seems to have been free alike from doubt and idle speculation. "Like the high-born in all realms," as another has well said, "in the realms of faith she began life at a point where the few end, and which the many fail to reach."

"The winter of her birth," writes Miss Fiske, "was one of special religious interest in the community. As the mother folds her precious child in her arms, she says, 'I hear the birds of Paradise on the boughs of free grace singing redeeming love. My soul can join in the blessed song, and I rejoice to see the work of the Lord prosper in the hands of the blessed Redeemer.' So she who was to labor so faithfully in revivals was, as it were, prayed into the work by those believing parents. Of the children given to Aaron and Jemima Lyon, one had become the 'family treasure in heaven.' Little Ezra had been in the Saviour's arms six months when Mary was welcomed to the mountain home, as she expressed it, 'to feel in that family circle the sweetly chastening influence of a babe in heaven.' She ever carried this with her, as well as the influence of another scene, where there were sorrowing hearts and flowing tears, because death had come to the same home to take away the affectionate husband and the kindest of fathers."*

The hallowed influence of these two events was perpetuated by the incense that continued to ascend from

---

* "Recollections of Mary Lyon," p. 15.

the family altar. The power of prayer was very early impressed upon her. When her mother would sink to rest exhausted, after long tarrying in her closet, her older sister would whisper, "I think there is going to be an awakening." Her confidence in the efficacy of her mother's prayers appears in the constant appeals for them found in her home letters. Holyoke pupils of 1840 will never forget her emotion when she told them, "I have no longer a mother to pray for me and my dear pupils." Her estimate of her mother's influence is told in the words, "I am more indebted to my mother than to all others except my Maker."

At the age of ten, she received distinct impressions that were never lost; but the first remembered indications of renewing grace were of later date. Miss Fiske says in her "Recollections":—

"It was a beautiful Sabbath afternoon of May, 1816, in which Mary Lyon first said, with full heart, 'Abba, Father'; 'Jesus my Saviour.' Her home was in the humble cottage at the foot of the hill. She had that day, as was her wont, mingled with the worshipers in the little Baptist church at the Three Corners. Good old Elder Smith talked both morning and afternoon of the character and government of God. At the close of the last service, the silver-haired man rose to bless his flock. He gazed upon them for a moment with more than paternal interest, and then said with deep solemnity, 'Remember, my friends, it is a fearful thing and a very wicked thing, too, not to love such a God as I have told you about to-day.' The fatherly hand was raised; there was heard 'Grace, mercy, and peace be with you all'; and the congregation scattered. Mary took the 'wild winding way' to her home. She trod that way but slowly, for her heart was too full for haste. As she approached the dwelling, an inexpressible feeling of tenderness stole over her. She remembered a scene in the 'north room' thirteen years before, when, a little child of six years, she heard her dying father say with faltering voice: 'My dear children,

BIRTH-PLACE OF MARY LYON.
Buckland, Mass.

what shall I say to you? God bless you, my children!' and then he was parted from them to enter into the fullness of blessing. The never-to-be-forgotten prayers of her mother passed before her and she exclaimed, 'Why should I not be blessed of my parents' God?' and turned away from her home to the hill-top to be alone with her Father in heaven. She dwelt upon his wisdom, holiness, mercy, and justice till peace came to her troubled soul and she exclaimed, 'O God, thy ways are perfect; be thou my Father and the guide of my youth, my everlasting portion.' Her heart now melted in love to him who had reconciled her to his Father and her Father. She looked upon the far off mountains in all their grandeur, on the deep valleys with the widely extended plains and the smiling villages below, and then thought of the kingdoms of the world, and, to use her own words, 'longed to lay them all at the feet of him who had redeemed' her. Twelve years afterward she wrote, 'I remember that moment as though it were but yesterday.'"

The desire which she thus felt and expressed at the age of nineteen was to become the inspiration of her life, but unbounded thirst for knowledge deferred for a time a life of consecration. At Byfield five years later, the self-classification which Mr. Emerson desired of his pupils brought her to face the question of her personal relations to God. She was greatly agitated, for though the friends of Christ were her chosen friends, she had not consciously classed herself among them. She did so then only after much deliberation, and with fear and trembling; but she was grateful ever after that she had been called to meet this test. Its salutary effect upon herself led to a carefully guarded use of the same measure in her own schools.

In March, 1822, she united with the Congregational church in Buckland. The inner life is known chiefly by the fruit it bore. She had a strong aversion to religious diaries. Indeed, she was always too busy planning, praying, and teaching, to keep a journal of her feelings. For several years, occasional remarks in her correspondence show that in assurance of hope and in special Christian activity, her spiritual growth was gradual. From the first a reverent regard for the Bible led her to improve every opportunity to show how its truths are indorsed by history or natural science. Enthusiasm over the beauty and sublimity of the Bible ripened into efforts to rouse the conscience and waken personal responsibility. By degrees it became her great aim to lead pupils "into the truth as it is in Jesus" that they in turn should do the same for others. She often said to an intimate friend, about 1828, "I think it very doubtful whether I ever see heaven myself, but I mean to do all in my power to prepare others for that blessed world." Having set herself to this work her own hopes grew clear, and her progress marked.

At the Centennial of the Buckland Church, in 1885, she was described as the most earnest Christian worker ever connected with it. The historian adds: "In all her later schools here, she labored first and most for the conversion of her scholars. The result was that through those scholars, revivals were carried to the towns around." The character of the school was so well understood that it is said when ministers in the sanctuary prayed for colleges, they prayed also for the school at Buckland. At Derry and Ipswich the dews of divine grace were almost constantly descending. In an address after her death President Hitchcock said: "A blessed result of her elevated piety was the almost constant presence in the schools which she taught of that divine influence which renews the heart. She lived to witness nearly thirty special revivals and not less than eleven in the twelve years of her new seminary." During the first six years not a graduate, and one year not a pupil, was left in the school without a hope in Christ. Another year, only three. In twelve years there were sixteen hundred pupils, and more than four hundred and sixty hopeful conversions.

By a remarkable coincidence Pliny Fiske, Jonas King, and Levi Parsons were born within forty days of each other and within twelve miles of the place where Mary Lyon was born less than five years afterward. Mary Lyon never crossed the sea, but her pupils are in every corner of the earth. Her first interest in foreign missions began in childhood with hearing of Carey, of Mills and his associates; it increased with the sailing of the first missionaries; it grew with the growth of the American Board, whose history from the first she eagerly followed. We have seen that her first Christian desires were to lay the kingdoms of the world at the feet of the Redeemer. New inspiration in this direction was gained at Byfield. On returning to her home, she organized the first missionary society in Buckland. With characteristic zeal she visited, in person or by proxy, every house in town, canvassing for members and for materials for work, letting down bars or climbing stone walls in order to reach more speedily the remoter dwellings. Over sixty children were enlisted. The socks they knit were exchanged for shoes and for cotton which their mothers wove into sheeting. In due time a box of socks, shoes, and twenty pairs of sheets was sent to the American Board. In these days of multiplied organizations for women and children it is of interest to note that it was nearly fifty years after this, and twenty years after her death, that the Woman's Board of Missions was formed, yet Mrs. Bowker, its president, attributes much of the present widespread interest to Mary Lyon, her teacher at Ipswich, and ascribes to the same person the beginning of her own interest in missions.

As a teacher, she sought by well planned efforts, not merely the conversion of her pupils, but their enlistment for the salvation of the world. The motives to which she appealed were Christian sympathy and a sense of personal responsibility. She labored with rare success to launch her pupils on a voluntary course of steadfast self-denial. In gifts to the Lord her own example led the way. The income of Mount Holyoke Seminary was the Lord's money. She would never accept from it more than a salary of two hundred dollars and a home within its walls. But from that salary she was not content to give a tithe. For several years before her death nearly one-half found its way into the treasury of the Lord. By her will she left to the American Board, in reversion, property exceeding two thousand dollars in value. Her school caught so much of the spirit of their head that in the last seven years of her life the amount of their contributions was nearly seven thousand dollars. That given by the pupils was taken from their allowance for dress and amusements.

Seventeen, at least, who had been under her instruction before she left Ipswich, became foreign missionaries. To these were added thirty-six of her Holyoke pupils, of whom two were associate principals, and seven others teachers at the seminary. With one exception each senior class for the first fifteen years had one or more representatives in the foreign field. Twelve other pupils of the first twelve years became teachers among the Indians in our own country. Of these forty-eight, nineteen did not finish the seminary course. Those who became wives of home missionaries, or teachers at the West and South, are numbered by hundreds.

A notice from the pen of Hannah Lyman published soon after her death, March 5, 1849, contains the following paragraph:—

"Is she missed? Scarcely a state in the American Union but contains those she trained. Long ere this, amid the hunting grounds of the Sioux and the villages of the Cherokees, the tear of the missionary has wet the page which has told of Miss Lyon's departure. The Sandwich Islander will ask why his white teacher's eye is dim as she reads her American letters. The swarthy African will lament with his sorrowing guide who cries, 'Help, Lord! for the godly ceaseth.' The cinnamon groves of Ceylon and the palm trees of India overshadow her early deceased missionary pupils, while those left to bear the burden and heat of the day will wail the saint whose prayers and letters they prized so much. Among the Nestorians of Persia and at the base of Mount Olympus will her name be breathed softly, as the household name of one whom God hath taken."

Another wrote: "Mary Lyon was one of the great spiritual teachers of the world. She possessed that very rare power of waking to activity the moral and spiritual nature. It has been said of her contemporary, Dr. Arnold, that he was the inspirer of more noble lives than any other man of the age. Dr. Arnold did a great work, but Mary Lyon's was greater. He came as a worker into an established order of things, and vitalized it with his own consecrated spirit. Mary Lyon created a new order of things, devised a system for developing dormant powers, and worked out her noble plan through evil and through good report."

Her life has not inaptly been called an epic. Rising from obscurity, she worked her way upward without retrogression. Advancing novel opinions, adopting original methods, she arrested public attention, stirred the hearts of Christians, and encountered inevitable opposition, yet kept straight onward, turning even defeat into a larger success. She knew her own defects and how to select her assistants. Her enthusiasm was contagious, her example inspiring. She knew what she undertook. The cost was counted. Plans were laid with deliberation. Obstacles were weighed, and either removed or overcome. The wise confessed her superior wisdom. Returning only blessing for reproach she won friends from opposers, and ultimately outlived the prejudice of all who came to appreciate her motive, and her work. Her sympathetic heart was great by nature; it expanded by consecration to Christ. By degrees all her magnificent capabilities were brought under the control of sympathy with his work.

In the words of Rev. Dr. Laurie, her last pastor: "Her most marked characteristic was not her love, embracing all, and yet loving each one as though alone. It was not untiring perseverance, or an executive power, seldom equaled. It was Christ formed in her and working by her. That one fact was the source of all her excellence. There has been but one incarnation. Only once has the Word become flesh and dwelt among us. But through union to him, that image of God in which Adam was created, is restored to his apostate children, and I never knew it so perfect as in the founder of this seminary. Hers was not a religion of human resolutions made prosperous, or help granted to human endeavor, but it was a being cut off from her own root and grafted into Christ, and so she bore much fruit. Lives like hers teach us that just as Christ once wrought miracles through the members of that body that was nailed to the cross, so now also he works miracles of grace through the Church which is his body, in such a sense that its members can say, 'I live, and yet no longer I, but Christ liveth in me,' and so we can say of Mary Lyon, and those who like her appreciate this relation to Christ. She wrought, and yet it was not herself, but Jesus Christ wrought his work through her."

CHAPTER III.

MOUNT HOLYOKE SEMINARY ANTICIPATED.

1818–1834.

THE history of Mount Holyoke Seminary reaches backward through the school at Derry and Ipswich, for they were essentially one, to that of Rev. Joseph Emerson at Byfield.

Before opening his seminary in 1818, Mr. Emerson had been a tutor in Harvard College and a pastor in Beverly. Besides doing other literary work, he had published a course of lectures on the millennium, a theme in which he took great interest. Among the influences to be concentrated for the hastening on of that era, he gave a high place to the power of woman. He regarded her chiefly as an educator of the race, since it is from her that children learn to walk, to talk, to think, to pray; and looked for the world's salvation mainly through her enlightened and sanctified instrumentality. He saw that the education within her reach was superficial and defective, and in a negative sense atheistic, and set himself to reform and elevate it; to substitute the substantial for the showy, the Christian for the worldly. His views were far in advance of his time. The former tutor believed in educating young women on the same broad and thorough principles as young men, and laid out a three years' course of study. Believing also in the joint culture of the mind and heart in order to obtain the best results from either, he gave the Bible the first place in the course. The formation of character, mental and moral, was his chief aim. He combined equal skill in stimulating pupils to think, and in leading them to feel individual responsibility. The principle he inculcated as pastor and as teacher was that of doing "the greatest good to the greatest number, for the longest time." Ann Hasseltine, the first American woman to be asked to become a foreign missionary, was a sister of Mrs. Emerson, and was much in their family. In the face of general disapprobation Mr. Emerson steadily encouraged her to go to India, and but for his efforts it is said to be doubtful whether she would have gone. Her decision undoubtedly influenced that of her friend, Harriet Atwood, who was asked the same question soon after. But Ann Hasseltine Judson and Harriet Atwood Newell were not the only ones who went forth from his influence to be life-long servants of Jesus Christ.

By training his pupils to investigate for themselves the principles of language, mathematics, mental and moral philosophy, he aimed not merely to secure a high standard of mental discipline in his own school, but to train teachers who should ultimately fill the land with similar schools. He had the co-operation of many friends, but met with opposition from others who doubted woman's ability to attain such results, or their practical utility if attained.

Among his pupils who put the question of ability beyond all doubt, were two who first met at Byfield in 1821,—Zilpah P. Grant, afterwards Mrs. William B. Banister, of Newburyport, and Mary Lyon. The latter at twenty-four had been seven years a teacher in the old brown or new red school-houses of Franklin county. The former at twenty-seven, twelve years from her first school in a log cabin near her Connecticut home, had just become Mr. Emerson's assistant. Her previous teaching had been marked with signal ability. Two years with Mr. Emerson as pupil and teacher gave her increase of knowledge, wisdom, and power. Mr. Emerson said of her: "Miss Grant has done more than any other young lady to raise my seminary."

To Mary Lyon, education became invested at Byfield with new meaning. Under God, she gave the credit to

Mr. Emerson. What he was to her through life is intimated in her later way of referring to him as "my dear teacher, now in heaven." In an address at the dedication of his seminary, in Saugus, January 15, 1822, Mr. Emerson predicted a time when higher institutions for the education of young women would be counted as needful as colleges for young men, and added,—"but when such an institution shall be built, by whom it shall be founded, and by whom taught, is yet for Providence to determine. Possibly some of our children may enjoy its advantages." The fruit ripened sooner than he expected. That same year Mr. Jacob Adams of Derry, New Hampshire, left the first bequest ever made exclusively for the academic education of girls. It led to the incorporation of the Adams Female Academy, in June, 1823. When a committee of the trustees were looking for a principal, they overheard the question, "How would Miss Grant do?" and the whispered response, "Brother Emerson cannot spare her." But when the matter came before Mr. Emerson, he saw in it a possible opening for his ideal institution, and though sorry to lose Miss Grant, he yet bade her God speed, saying: "If you can put into operation on right principles a permanent seminary for young ladies, you may well afford to lay down your life when you have done it."

Mary Lyon had gone from Byfield full of enthusiasm for the same great principles and was zealously planting them in the hearts of her pupils at Ashfield. When the good news of the incorporation of an academy for girls reached her, she at once hailed it "as an eminent means of doing good." The secret wish that her beloved Miss Grant might have charge of it, dismissed at first as a romantic idea, was soon revived by letters and a visit from Miss Grant herself, who laid before her a well digested plan for the proposed school and sought her aid in its execution. The plan pleased her. Its aim, like Mr. Emerson's, was to develop the mind and heart by a systematic course of study in which the

Bible should hold the first place; the completion of the course to be honored by a testimonial corresponding to a college diploma.

In April, 1824, the two friends opened the new academy. Of their sixty pupils, six were sufficiently advanced to be able, the succeeding November, to receive their testimonials, the earliest so far as known ever publicly conferred on young women. After four years of success at Derry a majority of the trustees became so dissatisfied with the prominence given to religious instruction that Miss Grant and Miss Lyon removed to Ipswich, Massachusetts. Most of their pupils followed them. There the number increased to nearly two hundred, and the course of study in both science and literature was extended as fast as public opinion would sanction. Assistant teachers were added as needed, the ratio of teachers to pupils being about one to twelve.

As the school in Derry was not in session from November to April, Miss Lyon opened in her native town a school on a similar plan to that at Derry, for teachers and others. Two winters in Buckland were followed by two in Ashfield, and these by two again in Buckland. Introducing at first only elementary studies, she added higher branches year by year and employed assistants as students and classes increased. Her work was not unappreciated. She came to be known and honored in a large circle of towns. More than half of the days of one term found visitors in the school, most of them from out of town. So eager was the demand of the public schools for teachers of her training that "committee men" were chosen in November instead of March as had been customary, that they might be ready to engage the best candidates. She was repeatedly urged to continue her work through the summer. During the sixth winter Rev. Dr. Packard of Shelburne was formally delegated to express the wish of the Franklin Association of ministers that she would remain permanently in that region; and funds were subscribed for a building.

But her summers had been pledged to Miss Grant, who in her delicate health regarded her co-operation as indispensable. While admitting that a school at Buckland "would be composed of more substantial materials than are generally found in seaports," and that Miss Lyon's sphere of usefulness there was exceptionally large, Miss Grant urged that more good would be accomplished by their continued union than by a separation, and desired her to remain at Ipswich in winter as well as summer. An attack of fever in August and another illness in February had made it plain that to labor in two fields so far apart was too much for even Miss Lyon's energy or greatest usefulness. When it was found that she would not leave Miss Grant, an attempt was made by the same ministerial association to induce Miss Grant to remove to Franklin county. Instead of that Miss Lyon gave the next four years wholly to Ipswich. This she did in the hope that they might establish on permanent foundations that system which had taken definite form under their united counsels. The permanent institution on right principles of which Mr. Emerson had talked so much, and which they were not permitted to establish in Derry, they hoped might grow up at Ipswich.

For a time Miss Lyon had been content with present opportunities, and had often said, "Never mind the brick and mortar, only let us have living minds to work upon." But at length the idea of permanence took as full possession of her as of Miss Grant, and both labored earnestly for the endowment of Ipswich Seminary. It was their habit to care for their pupils in and out of school as if they were their own daughters. To find boarding places and secure suitable arrangements for them in private families consumed too much of their time and strength. In a joint letter addressed to the trustees February, 1831, they set forth the need of a building for a boarding home in addition to one for instruction, and also of a library and laboratory with apparatus, urging that it would be no less difficult to sustain a seminary for young women without these appliances than a college for young men. This letter resulted in the appointment of a board of prospective trustees who, according to Miss Lyon's biographer, "held several meetings, passed sundry resolutions, and made many inquiries." Friends of the Ipswich teachers pledged one-half the needed sum, but the public was so apathetic that the failure of the project was anticipated. This led Miss Lyon, during Miss Grant's protracted absence in 1832, to prepare at the suggestion of the board, a prospectus of a new institution—The New England Seminary for Teachers—in which the superior advantages of a permanent over a private school were made the basis of an appeal for aid, the question of location being left for later consideration.

In November of the same year, attractive school buildings in Amherst were for sale. President Humphrey and Professor Hitchcock of the college urged that they should be secured for the proposed seminary. The Franklin Association had not forgotten Miss Lyon nor their appreciation of her work. A joint committee from that body and from the prospective trustees met in Amherst in April, 1833, and appointed a committee to call a meeting in Boston for further deliberation. So few came to the meeting that it adjourned, and the adjourned meeting utterly failed. The Ipswich board of prospective trustees dissolved, and the whole matter seemed to be at an end. But these efforts had not been useless. Attention had been drawn to the cause; a few had recognized its importance and saw that success would involve labor and sacrifice.

The principals at Ipswich patiently continued their work of training young women for Christian service, but without the least abatement of interest in their plan. It was a good plan and for a worthy cause. They were sure it would succeed, some day if not in theirs, somewhere if not in Ipswich. Their system had been tested. It made useful women, good mothers, and good teachers. In 1835, seven years from their beginning in

Ipswich, thirteen missionaries of the American Board, fifty-three teachers in the West and South, and three hundred teachers in New England, New York, and New Jersey, had gone forth from their school. Their success brought applications for teachers from nearly every state and territory in the Union. They saw the fruits of Ipswich training in the homes of the young wives and mothers they visited. Miss Lyon had tested the same methods at Buckland so far as her interrupted opportunities allowed, and the results were no less gratifying. Some of those Buckland pupils had become teachers in the Ipswich Seminary; two of them, Julia (Brooks) Spaulding and Abigail (Tenney) Smith, were already missionaries in the Sandwich Islands. Could she have read the future she would have seen Mary Billings in Madura as Mrs. R. O. Dwight and afterward in Madras as Mrs. Myron Winslow; Mary Grant, the first Mrs. Burgess, in India also; Mary Ann Longley, Mrs. Stephen Riggs, among the Dakotas; Mary A. White, teaching the boatmen on Lake Michigan; and many others, identified with the Lord's work at home or abroad.

With both principals the desire constantly deepened that the system producing such results should have a more permanent basis and larger advantages than were possible in a private school. "We had hoped," wrote Miss Grant, "that if an endowment were obtained the expenses might be somewhat less than in any existing institution, though this had never been presented as a prominent object." Just here lay a separate reason why Miss Lyon, though greatly prizing its opportunities, had not been satisfied with her field of labor at Ipswich. She fully shared Miss Grant's wish to bring together the young ladies of the higher and middle classes of society for their mutual benefit; but their pupils were mainly from among the more wealthy, and she knew of many with equal or greater aptness for learning and desire to be useful, who would value such privileges more than silver or gold, but who could not enjoy them for want of means. "My thoughts have turned," she wrote, "not to the higher, not to the poorer, but to the middle classes, which contain the main-springs and main wheels which are to move the world." "My heart has yearned over the young women in the common walks of life, till it has sometimes seemed as though a fire were shut up in my bones." On their account she had long wished to see the expenses at Ipswich reduced one-third or one-half. The failure to obtain the desired endowment there drove her to a closer study of the problem and to devising new measures for solving it. She noticed that while it was not a recommendation to a college to be expensive, there was a prevalent feeling that education for young women must be costly. She knew that at one of the most prominent schools in New England three hundred dollars had not been enough to support one young lady for one term, while one-third of that amount would pay the board and tuition of a pupil at Ipswich for the year of three terms. She remembered how the Buckland people in a winter when she made her charge for tuition only one shilling per week, responded by boarding her pupils for five shillings per week, thus securing to them for fourteen dollars, the board and tuition for fourteen weeks. She believed that a similar interest shared by teachers and patrons elsewhere would kindle a like liberality. How to awaken the interest became her study.

The argument drawn from the superior advantages of permanent institutions had not taken effect. She would try new arguments. The effort to enlist the wealthy and eminent had failed. She would try another class. She called to mind the way in which colleges were founded. Having been a pupil of Amherst Academy, she had watched with special interest the origin and growth of Amherst College. "Its funds were collected," she wrote Miss Grant, "not from the rich, but from liberal Christians in common life. At the commencement of that enterprise the prospect was held out that it would be a college of high standing where the expenses would be low, and that it would be accessible to all. This was the main-spring without which it is doubtful whether it would have been possible to raise the funds. I am inclined to think that something of this kind may be indispensable to our success."

The end she sought was not personal aggrandizement or emolument. It was personal in no sense. It concerned the welfare of the race. It had to do with the sex which molds in the nursery the coming men and women, which presides in the home and reigns in society; yet her zeal was not championship for woman. Her object was not the benefit of woman as woman, but the good of the world through woman. It was a benevolent object, and naturally led to a study of the benevolent operations of the day. Could not a seminary for the training of young women for Christ and the world, she asked, be founded and conducted on the same principles as benevolent organizations? Their treasuries were filled more by the smaller gifts of the many than by the larger donations of the few. The greater the number of actors in any work, the more widespread the interest. Again, those who gave themselves to a benevolent cause should and did expect for their services a mere support rather than ample compensation. Were there not teachers of a like spirit? She believed there were.

In seeking for further lessening of expenses, the idea of having the pupils contribute to this end with their own hands, suggested itself to this broad-minded, practical woman. "I have no faith," she wrote, "in any of the schemes of manual labor by which it is supposed that girls can support themselves at school. I should expect anything of that kind would become an expense rather than an income; but I am confident that arrangements can be made by which a family of young ladies can do the housework of the family, and without any sacrifice of refinement or loss in the acquisition of knowledge."

The idea was not to earn money but to lessen outlay; not to defray but to diminish expenses. She was not planning for idlers nor for the helpless, but for those who were able and willing to help themselves. In her view the self-helpful were the most likely to be the useful. That each pupil should thus bear her part in promoting the general good would only be carrying out the same unselfish spirit counted upon in donors and teachers. And this feature might remove much of the prejudice of those who thought that the higher education would unfit young women for practical life. The more she thought of it the more reasons she saw in its favor, till the pecuniary advantages sank into comparative insignificance.

At length her plan for founding a new institution took definite shape with these features: buildings to be erected and furnished by voluntary gifts; teachers to receive comparatively low salaries; domestic work of the family to be done by the pupils;—each feature grounded in Christian benevolence, and all together greatly reducing the cost of education. Nowhere in the world had a system of thorough intellectual training combined with careful religious culture been made accessible to the class of young women most likely to be benefited by it and to use it for the good of the world. If it could be done, though on a small scale, she felt assured that success would enlist public interest and secure means for enlargement. A single example would be worth more than all that had been or could be said on the subject. One such institution would be followed by others. Thoroughly possessed by these ideas her soul was stirred within her to see them realized.

Again the question of leaving Miss Grant arose. In the nine years of the Derry-Ipswich school she had been absent from its sessions but two winters. For a year and a half during Miss Grant's absence, she had had the entire superintendence of the school. The assistant teachers had been acquiring experience and ability.

Was it still true, as it had appeared three years earlier, that they could do less for the cause of education separately than together? For many months the growing question had been suppressed. When it became a subject of correspondence with Miss Grant, their letters show at what sacrifice of feeling to both the decision to separate was reached. Miss Grant returned to her post in the spring of 1833, and they decided to continue together another year. Leaving Miss Lyon again in charge, Miss Grant spent the summer of 1834 in traveling, and in the autumn their official relation was dissolved. Their lives were consecrated to the same cause and they continued as long as Miss Lyon lived, to confer together as before. In the summer of 1835 Miss Lyon returned to take Miss Grant's place once more in her absence. Separation did not lessen their mutual regard and each had gladly pledged to the field of the other her interest and her aid.

The following extracts from Miss Lyon's letters indicate her spirit in leaving:—

"I have longed to be permitted to labor where the expenses would be less than they are here, so that more of our daughters could reap the fruits. Sometimes my heart has burned within me; and again I have bidden it be quiet; I have thought that if I could be released from all engagements, perhaps I might in time find some way for promoting this object."

"At length I have decided to close my connection with this institution in the hope of using my limited influence towards advancing the belief that young ladies' schools of an elevated character may be furnished at a very moderate expense. I have much stronger desires to do something towards establishing some general principles than to accomplish much myself. But I hope that Providence will open a door where I may labor directly in a school in behalf of this great cause, as I believe I can do more in this way than in any other."

"In this movement I have thought and felt much more about doing that which shall be for the honor of

Christ, and for the good of souls, than I ever did in any other step in my life. I want that you should pray for me, my dear mother, that I may be guided by wisdom from above, and that the Lord would bless me and make me a blessing. My daily prayer is, Lord, what wilt thou have me to do? If the Lord go not with me let me not go up hence."

"I am about to embark in a frail boat on a boisterous sea. I know not how I shall be tossed, nor to what port I shall be directed, but it is sweet in the midst of darkness to commit the whole to His guidance."

"The question of the expediency of devoting myself to this object in some place farther west has been several times mentioned to me. But considering that improvements in education seldom make any progress eastward, and that New England mind carries the day everywhere, my purpose to live and labor in New England has become fixed."

"I have no definite spot in view where I may spend the remnant of my strength in behalf of an object which for a long time has seemed to drink up my spirits; yet I never had a prospect of engaging in any work which seemed so directly the work of the Lord as this. The present path is plain. The future I can leave with Him who doeth all things well."

CHAPTER IV.

PREPARATIONS.

1834—1836.

WHEN Miss Lyon withdrew from labor at Ipswich it was with a distinct conception of the seminary she was to found. For six months she had distributed extensively a printed circular, addressed to the friends and patrons of Ipswich Seminary, containing the main features of her plan. She believed that the arguments which had commended the plan to her own understanding, if fairly presented, would convince many others.

On September 6, 1834, some days before the term closed, a few gentlemen of large views and larger hearts met by invitation in her private parlor to devise means for founding a permanent seminary according to her ideas. They appointed a committee to commence operations at once, with authority to act till the appointment of a permanent board of trustees. The committee consisted of Rev. Daniel Dana, D.D., of Newburyport, Rev. Theophilus Packard, D.D., of Shelburne, Rev. Edward Hitchcock, Professor in Amherst College, Rev. Joseph B. Felt of Hamilton, George W. Heard, Esq., of Ipswich, Gen. Asa Howland of Conway, and David Choate, Esq., of Essex.

The committee appointed at this quiet meeting scarcely known to twenty persons outside the room, supplied their own vacancies, and added to their number from time to time, Rev. Roswell Hawks of Cummington, Rev. William Tyler and William Bowdoin, Esq., of South Hadley Canal, Rev. John Todd and Rev. Joseph Penney, D.D., of Northampton, Rev. Joseph D. Condit of South Hadley, and Samuel Williston of Easthampton. Till a charter was obtained these men stood before the public as responsible agents for establishing the proposed seminary. Some of them became trustees and others resigned their places on the committee.

Within two months from the meeting in her parlor, Miss Lyon collected from women in Ipswich and vicinity nearly one thousand dollars for expenses of agencies and other preliminaries; January 8, 1835, the committee decided upon South Hadley as the location, provided the subscription there could be raised to eight thousand dollars; April 15, 1835, the seminary was named; February 10, 1836, the charter was granted; May 19th, the site selected; and October 3rd, the corner-stone laid. September 6, 1837, just three years from the meeting in her parlor, Miss Lyon wrote, "Our building is going on finely. The seal to everything is soon to be fixed. My head is full of closets, shelves, doors, sinks, tables, etc." November 8, 1837, the school opened.

These sentences contain much more than appears. That Mary Lyon by her own personal efforts should raise one thousand dollars in two months, speaks not only of enthusiastic zeal on her part, but of answering interest on the part of the donors. From the pupils of the seminary, she received a free-will offering of two hundred and sixty-nine dollars. A former pupil sent one hundred dollars from the far South, whither she had gone to teach. The ladies of Ipswich, grateful for the privilege, gave her four hundred and seventy-five dollars. The rest was given by women in the neighboring towns, before the seminary had a name or a place, and when there was no expectation that it would be located nearer than Worcester county or the Connecticut valley. But Miss Lyon was well known as a thorough teacher, a successful manager, and an honorable woman. The money was paid when it was solicited, and was the pledge of future success. She always called it the corner-stone of the institution. It was the first known attempt for advancing the education of woman by public benevolence and was a thorough committal to the object. Every dollar was well invested and brought a hundred per cent. When the general fund was to be raised, the story of the liberality of the Ipswich women was carried from town to town with an eloquence that stirred the hearts of others.

The pupils who had gone forth from Ipswich were representatives of the system she wished to perpetuate. Their well known character won favor for her plans. It was the testimony of herself and her agent that wherever they found those graduates they gained a readier access to the hearts of the people. What the Ipswich Seminary did for her in eastern Massachusetts the Buckland school did in the western portion of the state.

Yet it is difficult at the present day to appreciate the obstacles encountered by an enterprise so new, or the courage and persistence needed to overcome them. When Miss Lyon asked the *Boston Recorder* to publish some articles in her favor, the reply was that they would if she paid for them as for advertisements. But she had more than indifference to contend with. "Respectable periodicals," says Dr. Hitchcock, "were charged with sarcasm and enmity to her plans. So ungenerous were some of these attacks, that I volunteered in her behalf. I found her entirely unruffled. She did not object to the spirit or style of my defense, and I left it in her hands to be published if she thought best. But that is the last I ever heard of it." That she never destroyed his paper shows that she was not insensible to the kindness of her friend; but her only reply to such attacks was that of another builder, "I am doing a great work. I cannot come down." She was not indifferent. "No one can be more sensitive to such criticisms. I feel them keenly," she said, "but I receive them as a severe yet indispensable test of my character." There would be a brief struggle, then a smile and the gentle remark, "Well, we will go on."

Through Rev. Dr. Packard the enterprise was brought before the Massachusetts General Association at its meeting in Lee, 1834. A committee was appointed and a favorable report made. But when the minutes were read at the close of the meeting, such opposition appeared that a vote of reconsideration was passed and the recommendation erased. "Thus you see," wrote Dr. Packard, "that the measure has utterly failed. The hand of the Lord is in all this. Let this page of Divine Providence be attentively considered in relation to the subject." Attentive consideration did not prevent a renewal of the effort. Of the next meeting of that body Miss Lyon wrote: "Rev. Mr. Todd requested Rev. Morris White to bring the seminary before the Association. The subject was presented and a committee of five appointed to report on it. Mr. White was not on this committee, and probably not one who was known to have any interest in the object. In the committee one objected because he was a trustee of an academy. A second was much more opposed. Mr. M., a quiet, good man, would do nothing any way; Mr. C., a candid man, the youngest of the five, was favorable. Mr. White and Mr. C. sought to bring the matter to a close in some way not injurious to the object; so Mr. White told the committee that he only wished to secure some general resolves in its favor, but as they were opposed to it, he would withdraw the proposition. This brought them to terms and they reported three resolutions. The first was in favor of Christian education among women; the second granted that sufficient effort had not been made; and the third recommended Mount Holyoke Seminary or any other institution designed to effect a similar object; and they were passed without opposition."

That a woman with such aims could have other than selfish motives was too rare to be readily understood even by the best men of those times. Some who might otherwise approve the work feared it would soon end. When Miss Lyon was talking it over with Prof. Emerson of Andover, the brother of her revered teacher, we are told by Mrs. Haven: "My father asked 'who shall he be that cometh after the king? Will it not die with you?' 'No, it will not, we shall raise up our own teachers and it will go on,' she replied, immediately resuming the subject of her enthusiasm."

Thoroughly convinced of the worth and practicability of her cause, she did not fear to stand and act alone, patiently waiting for others to see the subject as she did, for she was certain that the object would finally commend itself to the good common sense of New England. Save for this faith, she could never have enlisted so many heads and hearts and hands as were needed to carry out her plan. Yet it was a severe trial to go forward in opposition to the opinions of some of the best and wisest. Each of the principal features of the plan was opposed. There were good men who had no faith in the success of appeals confined to the motive of benevolence. Even Miss Grant objected to low salaries for teachers, and thought the domestic feature unadvisable.

Miss Catharine Beecher wrote: "I fear you are starting wrong. It is the object of great plans to raise the profits of our profession. If this is not secured the profession will be forsaken by energy and talent and be the resort of the stupid and shiftless. It cannot be sustained by the missionary spirit. That will send forth ministers and missionaries, but rarely teachers. Therefore all plans that tend to sink the price of tuition will probably be discountenanced by the most liberal and expanded minds that are engaged in the enterprise."

Miss Beecher recommended setting a high price for tuition, with the understanding that all who needed should receive aid. Miss Grant and Miss Lyon had for some time been giving such help by gift or loan. It led to the formation in 1835 of the Society for the Education of Pupils in Ipswich Seminary—the first Education Society for women. But Miss Lyon preferred to bring tuition within reach of the largest number and to build on the broad educational principle of self-help.

To Miss Grant's objection Miss Lyon replied: "While the public are so little prepared to contribute liberally to an object like this, may it not be expedient that those who first enter the field as laborers should receive as a reward so little of 'filthy lucre' that they may be able to commend themselves to every man's conscience, even to those whose minds are narrow, and whose hearts are not much enlarged by Christian philanthropy? If such a course should be desirable at the commencement, how soon it would be no longer needful, time and experience alone can decide."

Her answer to Miss Beecher follows:—

I thank you for your interest in my plans, expressed in the sincere way of criticism on one point, yet I think you do not fully understand them. The terms high, low, and moderate tuition mean different things in different parts of the country. In speculating portions where wealth flows in as in a day, in some of the most prosperous mercantile and manufacturing places, and in the South where wealth is concentrated on large plantations, these terms are understood differently from what they generally are in New England. Its people tilling a sterile soil and uniting economy with prudence, are enabled by the slow gains of patient toil to provide comfortably for their children and send them to school in their own neighborhoods; to sustain the ordinances of the gospel, and to cast something into the treasury of the Lord in order to send the gospel to the heathen; to raise up ministers; to build up colleges and seminaries at the West; and to supply with the sacred ministry the destitute of our own land, who are less able or less willing than themselves.

Our plan is to place tuition at what will be regarded by the entire New England community as moderate tuition. Here let it be distinctly understood that we do not adopt this standard because we consider ourselves under any obligation to man so to do. Neither do we consider it necessary that other institutions should adopt the same standard, or that this institution should certainly abide by it evermore, though at present it is essential to our success.

I have not been alone in considering it important to establish a permanent seminary in New England for educating women to be teachers, with accommodations, apparatus, etc., somewhat like those for the other sex. Honorably to do this, from twenty to forty thousand dollars must be raised; and such a sum, raised for such an object, would form an era in education for woman. For years, Miss Grant and myself made continual efforts to accomplish this; but we failed.

I am convinced that there are but two ways to accomplish the object. First, to interest a few wealthy men to do the whole; second, to interest the whole New England community, beginning with the country population, and in time receiving the co-operation of the more wealthy in our cities. Each of these modes would have its advantages. The first could be done sooner and with comparatively little labor. The second requires more time and labor; but if accomplished, a salutary impression would be made on the whole of New England.

Having adopted this second course, we have been for some time as successful as we could expect. We have enlisted for the work. I have regarded it as a work for life. In laying our plans, we examined carefully every step. In the commencement of any great enterprise, the community are often unprepared to act upon the most important considerations, while moved by less important, but more tangible circumstances. During my long but fruitless efforts in connection with Miss Grant, I became convinced that the community were not prepared to appreciate the most important advantages of an institution thus endowed, such as its superior character and its permanency. I was also convinced that, to give the first impulse to this work, something more tangible must be presented, of real, though of less value, and that it must be made to stand out in bold relief. For this purpose, we have chosen the reduction of expenses as compared with other seminaries. Every step we take proves it a good selection. We carefully avoid all extravagant statements; indeed, we usually state only general facts, leaving each to make his own estimate and draw his own conclusions. There is an expectation that economy will be practiced in the establishment, that the funds, gathered by little and little, will be reserved for the good of the institution, and not for private emolument, and that there will be such a reduction of expenses as the nature of the case will allow. Here is our pledge, and we must redeem it. In doing this, the first object to be gained is good management in the boarding department. Let that be secured, and all else will be sure to follow. I do not expect to have the direct care of the boarding department, but I hope to secure the co-operation of those skilled in domestic economy, and disposed to use their skill faithfully. The department of instruction I expect to superintend myself; and it is essential to success in the boarding department that I should set an example of economy in my own. Otherwise, I cannot influence this point in other departments. I do not mean to ask any other one connected with the institution to make such sacrifices as I can cheerfully make. This may not be necessary for my successor, but it is necessary in my case, at least for a few years.

Again, we have held up the advantages of a teachers' seminary, with ample facilities for boarding and instruction, free of rent, of so superior a character that a supply of scholars could be secured without receiving the immature and ill prepared, who are always a tax on the time of teachers. We have shown that the same money will in this way do more to aid young women to qualify themselves to teach, than it would in our country academies. After these professions, shall we ask for higher tuition at the same time that we are asking for aid to carry forward our enterprise?

I feel confident that we must retain our plan for tuition, or abandon the enterprise. But we must not give up the work. To indulge even a fear as to our final success, would be to distrust the kindest Providence. While I do not consider ourselves under any obligation to man we are under solemn obligations to God to adopt this course. We are compelled by the principle of expediency, so beautifully exhibited in the precepts and practice of the apostle Paul. If any injury should result to the cause of education from our adopting this moderate standard of tuition, it will be as nothing compared with the great good to be accomplished; far less than the injurious results of the example of Paul, on the support of the gospel ministry, which results he so carefully guards against in the ninth chapter of First Corinthians.

I express myself with more confidence on this subject, because it has been with me, for two or three years, a matter of careful consideration; but further, because our indefatigable agent is of the same opinion, and he probably knows more of the views of the New England community on this point than any hundred others.

You speak of the importance of raising the compensation of teachers. In a list of motives for teaching, I should place first the great motive, which cannot be understood by the natural heart, "Love thy neighbor as thyself." On this list, though lower in rank, I have been accustomed to place pecuniary considerations. I am inclined to the opinion that this motive should fall lower on a list to be presented to ladies than to gentlemen, and that this is more in accordance with the system of the divine government. Let us cheerfully make all due concessions, where God has designed a difference in the situation of the sexes, while we plead constantly for the religious privileges of woman, for equal facilities for the improvement of her talents, and for the privilege of using all her talents in doing good!

At one time the Faculty of Amherst College supplied the pulpit of Rev. Melancthon G. Wheeler, of Conway, that he might go to eastern Massachusetts to introduce Miss Lyon to leading men of his acquaintance. He never saw discouragement in her but once. They had gone with high hopes to see a Doctor of Divinity of well known liberality and of unusually ample means. While the agent introduced the subject she remained at the hotel expecting at least to be invited to call. He not only failed to apprehend her object but made light of the whole plan, especially its domestic feature, in the presence of his daughters whom she had hoped to have among her first pupils, and then merely inquired where Miss Lyon was, and sent his respects to her, without inviting a call. On learning this, Miss Lyon buried her face in her hands and bowed her head on the table in keen disappointment. In a few moments the cloud passed and she rose saying, "If God wants me to succeed, I shall succeed. We will go on." It is due to that clergyman to say that he afterwards apologized for the manner of that day, and confessed that he ought to have had more faith in the undertaking of a Christian woman like Miss Lyon. It is but just to say that most of those who opposed her plans ultimately acknowledged their wisdom and gave her their aid. Many were friendly but incredulous. One wrote ten years later, "I remember when you explained your scheme to me I thought it excellent, only I was afraid it would prove like a wonderful machine Dr. Beecher used to tell us of, admirably contrived and admirably adjusted, but it had one fault,—it would not go. You see I was mistaken; it does go most beautifully. I rejoice in your success, and there is joy in heaven over it also."

So long and prayerfully had Miss Lyon surveyed the whole ground that she could not turn back without doing violence to the strongest convictions. In 1835 she wrote, "I have no doubt I am following the leadings of Providence. His dealings towards this new enterprise have been such as should lead me to trust wholly in the Lord. Every success has been from his hand, and every discouragement has been such that when good comes we feel constrained to say, 'This is the Lord's doing.' It seems to me more and more that this and similar institutions are a necessary part of the great system of means for the conversion of the world. The feeble efforts which I am allowed to put forth in co-operating with others to lay these foundations will probably do more for Christ after I am laid in the grave than all I may do in my life. It is a great privilege to labor for him in any place and in any circumstances he may direct, and a still greater privilege to lead others to do more than I can ever accomplish." "From some indications I expect trials in future such as I have never known. Sometimes I am almost ready to exclaim, 'When will the work of my feeble hands be done that I may go home?' But through the mercy of God these seasons are not frequent and do not continue long. Generally I feel that the dark cloud which hangs over the future is under the direction of Him who led his people by a pillar of cloud and of fire."

She longed for the sympathy of friends, but was willing to go on with only a very few. Friends were given her. If some failed, others took their places. The peculiar features of her plan became the means of her success.

After leaving Ipswich in 1834, Miss Lyon took up her abode for the winter in Amherst, attended some of the college lectures, and reviewed the natural sciences, to be the better prepared for future teaching. She improved every opportunity to talk of her project with intelligent people whom she met. Whenever there was a prospect of promoting her plans by her presence elsewhere she was sure to go. By desire of the committee she attended their meeting at Worcester to decide upon a location. The mercury was below zero in Amherst that January morning when she and Prof. Hitchcock took seats in the stage, three or four hours before sunrise, each wrapped in a buffalo robe. Andover, Worcester, Brookfield, and Northampton had been talked of, and Rev. Cyrus Mann had presented the advantages of Westminster; but South Deerfield, Sunderland, and South Hadley had each offered a handsome subscription to secure the seminary. Always preferring the central or western part of the state, Miss Lyon was intensely interested in the question but was satisfied to leave the decision with the committee. The location decided, a name must be found. Prof. Hitchcock had published in newspapers an outline of the proposed seminary and suggested for a name, "The

Pangynaskean Seminary," meaning the seminary in which all the powers of woman—physical, intellectual, and moral—were to be cultivated. He intended by the Greek term both to designate its leading features and to attract public attention. He succeeded. Other papers took up the phrase, and by their ridicule of the "whole-woman-making school" gave it free advertisement. But the name fell into disfavor, and when the subject was discussed in the committee, Dr. Todd said, "Call it Mount Holyoke; then the name will indicate the locality." His suggestion was at once adopted.

The Greek term had been the occasion of so much sarcastic remark that Miss Lyon's friends feared the enterprise would be injured. But those newspaper articles did far more good in a single case than all the harm they could do. They were read by a lady in Connecticut who had once been sent to a school where the instruction received and the money paid seemed almost in an inverse ratio, and she had had the good sense to discover it. A seminary of high order proposing to put expenses at cost roused her interest. She soon after became the wife of a prosperous business man of Boston, who had decided that all the increase of his property above needed expenses should not only be the Lord's, but should year by year be spent in his service. Riding in their carriage from Boston to Belchertown, this lady entertained her husband with an account of the projected seminary, and they agreed to give it a part of their surplus funds. As they drove into the town their attention was attracted by some unoccupied buildings. Learning that they were for sale, they sent a letter to Miss Lyon asking whether she could make use of them. Twice before, Miss Lyon had heard that name. Mr. Stoddard, of Boston, had told her, "He is just the man to carry forward your work"; and while she was praying over the matter Prof. B. B. Edwards, who had married one of her Buckland pupils, had recommended the same man as reliable for counsel and aid. When the letter came, thinking he might be the owner of the buildings and desirous to sell, she feared he would not look on her plans with favor. Yet she ventured to write him, requesting an interview. He replied by inviting her to his house. She used to tell her pupils years after with moistened eyes, "I cannot describe my feelings when I found myself at his door. Between the ringing of the bell and its response, I tried to roll all my care upon the Lord and be willing to receive not one encouraging word, if so my God would be most honored." As she unfolded her plans, watching intently every expression of her hearers, she saw that they listened with eager interest, but had then no knowledge of the way the Lord had prepared them for her visit. When by themselves the husband asked, "How much do you think I should give Miss Lyon?" The wife replied, "I thought perhaps you would give five hundred dollars." He was surprised at the answer, but rejoiced the friends of the cause by affixing that sum to his well known name. It was the first, but by no means was it the last five hundred dollars he gave. His interest in the seminary increased until it became his favorite work, occupying his thoughts as well as receiving his money, and filling scarcely a smaller place in his affections than in Miss Lyon's. "What I have given to Mount Holyoke Seminary," said he, "I consider the best investment I have ever made; there is no depreciation in the stock; it yields the largest dividends." But the time, influence, and sympathy which he and his wife gave to it were worth more than their thousands of silver and gold. From the day of that visit to the day of her death, Deacon Safford's house was Miss Lyon's home in Boston.

Rev. Dr. Packard was a connecting link between the school at Buckland and Mount Holyoke Seminary. We have seen him communicate to Miss Lyon the action of the Franklin Association in 1829. He was chairman of the general committee appointed in

Ipswich, September 6, 1834, and was the first agent employed. He was a friend in the infancy of the enterprise, when most needed, and gave much time and energy to devising ways for securing subscriptions. One plan was to have scholarships of two hundred and fifty dollars each, owned in shares, the owners being at liberty to send a pupil for fifteen dollars a year less than others. Miss Lyon saw great objections to this plan. After full conference, at a meeting in Ipswich in December at which Dr. Packard was present, the committee decided to depend on the free will offerings of an enlightened public.

Unable to devote himself to the work, Dr. Packard introduced to Miss Lyon a younger man, Rev. Roswell Hawks of Cummington, who had previously been a pastor in Peru. Like other New England pastors he took young men into his family and fitted them for college; Cummington has been noted for the prominent men reared there, some of whom, Henry L. Dawes, William C. Otis, Eli A. Hubbard, and W. W. Mitchell, were his pupils. He was instrumental in establishing two schools in Cummington, both of a high order, but neither on a permanent basis. His daughters often heard him say, "While so much is being done for young men, there is not an endowed seminary in the land for our daughters." For years he had been studying how to secure greater advantages to them. It was his favorite theory that as woman was the occasion of the fall, she is to bring back into society, the family, and the church, those influences which prepare the way for the coming of the Son of God.

"At a meeting in Boston in May, 1834," his daughter writes, "he met Dr. Packard and said, 'I want to confer with you in regard to a plan for the education of the daughters of our land.' Dr. Packard replied, 'If you have that in view you should see Miss Lyon of Ipswich, who is here for the same purpose.' She was not unknown to him, for he had sent a daughter to her school in Buckland. They met; Mr. Hawks, entering heartily into Miss Lyon's project, declared himself ready to do anything in his power to forward it, and from that day onward, no person was a more patient listener to her plans or a more sincere co-worker in their execution. She came to have such confidence in his judgment that she would undertake no important measure without first consulting him, and neither would adopt a course the other did not favor. Each had the same objections to Dr. Packard's proprietary plan for raising funds." The following account given by one of his daughters refers to a meeting before the summer of 1835: "Miss Lyon and father met at the house of Dr. Hitchcock in Amherst, to go to Boston to a meeting of the committee when foundation principles were to be settled. Long before dawn Mrs. Hitchcock had served their breakfast and they were waiting for the stage coach, but no coach came. They had been forgotten, and as there was no other conveyance they had to wait till the next morning. They were two days on the way and meantime the important meeting was in progress. Late in the afternoon the two weary, anxious travelers entered the room to be told that principles had been adopted, plans formed, and the meeting was about to adjourn. Had the adjournment taken place, the Holyoke of to-day had never existed. The conclusions reached related to two subjects; one involved the principle to be followed in soliciting funds, the other was a question whether Miss Lyon's name alone would be regarded by the public a sufficient guaranty for the success of an enterprise needing so much money. It seemed to the committee that Miss Grant's name also was needed, and therefore that the plan must be modified to meet her views, though it would require radical changes. But when Miss Lyon explained her plans, both decisions were reversed and Mount Holyoke Seminary was given to the world. Not being then a member of the committee, father withdrew from the room with the request that he might be notified when the subject of agencies came up. A long time he waited. Then Deacon Safford's genial face appeared. 'Go home to tea with me,' he said. Father replied, 'I am waiting to speak on the subject of agencies.' 'Too late,' said Deacon Safford, 'the work is done, the meeting adjourned, and you are appointed sole agent, for you have faith that money can be raised, and that it can be done on the benevolent principle, and we have not.' 'It was for that reason I intended to decline,' was the answer."

His people were unwilling to relinquish him. Some of them ridiculed the scheme of a seminary for women. Some thought he must certainly be of unsound mind. One of the most influential told him he had mistaken his duty; that he could do a thousand-fold more good by laboring for souls in his parish than by establishing that school. But he was full of faith that the Lord had a greater work for him to do and told his people that if by his remaining with them every soul in Cummington would be converted, he would not remain, for he felt sure that the establishment of a seminary for the education of women would be a far greater work. So strong was the feeling of his people that they declined to contribute for the seminary, claiming that in giving their pastor they had done more than any other town.

His whole heart was enlisted and he entered without delay upon the task of raising funds for the project called chimerical by some and by some even wicked. How his faith triumphed is attested by the very walls of the seminary; but the fatigue, cold, and hunger he endured in his journeyings, that the funds of the seminary should not be lessened for his needs, was then known only to God and himself; in after years it became known to his family.

Of the noble women who gathered round Miss Lyon, none is more worthy of mention than his wife, Mrs. Eliza (Green) Hawks, who took on herself unwonted burdens and met cheerfully many a sacrifice that he might give all his time to the cause equally dear to both. That was a joyful hour in 1842 when they saw their three daughters receive the diploma of Mount Holyoke Seminary.

Mr. Hawks did a similar work for Lake Erie Seminary in Painesville, Ohio, but his love for the first child of his toils and prayers was strong to the end, and by his request his body sleeps in the cemetery within sight of its walls, and within sound of its bells.

In June, 1835, the committee voted to invite the ladies of the Connecticut valley in Massachusetts, to raise one thousand dollars, and addressed a circular to the Christian public from which these extracts are taken:—

After much deliberation, prayer, and correspondence, the friends of Christ have determined to establish a school for the daughters of the church, whose object shall be to fit them for the highest usefulness. The justly celebrated school at Ipswich embraces most of the features which we desire in this.

1. It is designed to be permanent ; to be under the guardianship of those awake to all the interests of the church. It will not depend on the life of a particular teacher, but like our colleges be a perpetual blessing.

2. It is to be based entirely on Christian principles, and while furnished with teachers of the highest character, and with every advantage that the state of education in this country will allow, its brightest feature will be that it is a school for Christ.

3. It is located at South Hadley, Massachusetts, on the banks of the Connecticut, at the foot of Mount Holyoke, in the center of New England, easy of access from all quarters, and in the midst of the most delightful scenery.

4. The buildings are to accommodate two hundred young ladies.

5. It is designed to cultivate the missionary spirit among its pupils; no romantic idea of moving in some high sphere, but the feeling that they should live for God wherever he may appoint their lot.

6. The seminary is to have a library and apparatus equal to its wants; and such internal arrangements that its pupils may practice those habits of domestic economy that are appropriate to the sex, and without which all other parts of education are too expensive.

7. The seminary is to be placed on such a pecuniary basis that all its advantages may be within the reach of those in the common walks of life. Indeed it is this class principally, who are the glory of our nation, that we seek to help. The wealthy can provide for themselves; and though we expect to offer advantages which even they cannot now command, yet it is not for their sakes that we erect this seminary. We intend it to be like our colleges, so valuable that the rich will be glad to attend it, and so economical that people in moderate circumstances

may be equally accommodated. We expect that distinctions founded on wealth will find no place within its walls any more than at the table of Jesus Christ.

8. In order to establish it, the committee believe that not less than thirty thousand dollars is needed. Everything is to be done as economically as possible, yet the materials and work should be the best of their kind. Of this sum, South Hadley has pledged eight thousand dollars, which with other subscriptions makes about one-third of the sum required.

The object and plan have been in many respects grossly misrepresented—probably through ignorance. But wherever they have been understood there has been but one voice, and that in their favor.

We have daughters who would gladly become teachers, and go anywhere to do good—were they only prepared. We have a population of millions calling loudly for instruction. The spirit of enterprise is such that we cannot induce young men to become teachers. We must look to the other sex for a supply. To obtain it this plan has been long maturing. It commends itself and will succeed, for it is the offspring of prayer, and if any were ever actuated by pure motives, we believe those are who have been praying this seminary into existence. We commend it, dear brethren and friends, to your sympathies, prayers, and charities.

Signed, JOHN TODD,
JOSEPH PENNEY, } *Committee to address the public.*
ROSWELL HAWKS,

The following extracts are from a pamphlet by Miss Lyon, printed in September, 1835, for the benefit of candidates for admission.

"It is desirable that friends should carefully consider the design of this institution before influencing any to avail themselves of its privileges. Its main features are an elevated standard of science, literature, and refinement; and a moderate standard of expense: all to be controlled by the spirit of the gospel.

"Its object is to meet public and not private wants; to provide not for individuals only, but for our country, and for the world, by enlisting the talents of our most gifted daughters. Some may be wealthy; some may be fitted for the service by an answer to Agur's prayer; others may struggle under the pressure of straitened means; but we hope the desire to do good will be the chief motive, bringing together congenial souls. Unlike many institutions of charity this does not provide for the relief of individual want, nor directly for the instruction of the ignorant and degraded."

She could not describe its literary standard by comparing it with that of established institutions of the kind everywhere known, as one could do in founding a new college for men. There was no other school to which she could point as an example in this respect when she added: "It is to take the literary standard of Ipswich Seminary, allowing for continual progress, just as that institution has been advancing from year to year. It is to adopt the same high standard of mental discipline and thorough investigation, and the same systematic course of solid studies." An outline of that course was followed by the remark, "That it may accomplish the most good, it is designed for an older class of young ladies, and it is desirable that they should advance in study as far as possible before entering."

Mr. Hawks was diligently soliciting funds. Miss Lyon often went with him from town to town, although at great cost of feeling, for she knew her motives were misconstrued. July 24, 1835, she wrote Miss Grant: "The more we seek to draw the public to aid us, the more perplexing will be our work; but we shall not shrink from this if we can thus lay a foundation for our successors to labor abundantly for Christ." She was urged to rely less on personal efforts and more on her pen and the agent. But pen and agent were already doing their utmost. It was the Lord's cause and she was willing to make herself of no reputation if necessary for its advancement. Her pupils recognize the motto she was following, "What ought to be done can be done, and you are the one to do it if no one else is ready." Her persuasive eloquence was remarkably successful, yet long afterward she said that if she had known how much she would have to suffer in these efforts, she might never have made them, adding joyfully, "but perhaps I should not have prized sufficiently my present opportunities, had I not passed through that trial."

For the hundreds of miles she traveled between September, 1834, and November, 1837, the funds of the seminary were never drawn upon, nor for the postage of her large correspondence—when the rates for single letters were from six or ten to twenty-five cents each. From her own purse she expended for the cause from twelve to fourteen hundred dollars.

She had learned not to look to the wealthy for the most efficient aid. In Conway she visited Deacon Joseph Avery, of whom she wrote January 11, 1836: "During the past twenty years he has probably given more to benevolent objects in proportion to his property and family than any other man in New England. I was delighted with the godly simplicity, well balanced views and systematic benevolence of the family."

For several years this good man gave the seminary substantial aid from his rock bound farm. Like Miss Lyon, he could not bear to see a cent of the sacred funds of the seminary go for naught. At one time when an artist's plan had been purchased which did not prove available, he paid the bill in addition to his large subscription. This kind deed, at a time when her own purse was poorly able to bear another draft, Miss Lyon never forgot. When he could give no more in money toward the erection of the first building, he came to South Hadley and gave the labor of his hands day after day. Deacon Avery was a progressive man. After the addition to the building in 1842, when the trustees with some anxiety faced the question whether the new seminary hall should be carpeted, he was one of the first to approve the outlay, saying, "The times demand it. The education of the world is being carried on here." Though without college training he was not unlearned. During an algebra examination in the seminary, another trustee said, "I suppose this is all Greek to you, Deacon Avery." "Not quite," he replied; "when my boys were studying algebra and geometry I studied with them."

In Conway also the Misses Maynard pledged their prayers and efforts. How this came about is told by a niece who lived in Ashfield near Miss Lyon's mother: "Miss Lyon came from her mother's to our house one day with a plan to visit my aunts in Conway, and I was allowed to go with her. I can never forget that ride nor how full of enthusiasm she was in the cause of education for women. My aunts each promised her one hundred dollars, and though they soon after lost their property they were unwilling to lose the privilege of helping Miss Lyon to the full extent of their pledge. I remember also how her mother used to come in and talk with my mother about her discouragements. One day after a long talk, she threw herself back in her chair saying, "But Mary will not give up. She just walks the floor and says over and over again when all is so dark, 'Commit thy way unto the Lord, trust also in him and he shall bring it to pass.' Women must be educated—they must be!"

In a cabinet in Williston Hall are to be seen the coins referred to by a Holyoke pupil in the following paragraph. "Well do I remember standing with Miss Lyon by her open drawer, as she took up several silver dollars bearing the traces of fire. Her eye kindled as she said, 'These were among the first contributions to our seminary. They were given by two sisters whose house was burned after they had subscribed one hundred dollars each. We felt that they were released from obligation, but they earned the money with their own hands and paid the whole. These dollars gathered from the embers were part of their gift. I replaced them with my own money and kept them as a memento of their faithfulness and of God's goodness to the seminary.'" With such money was the institution built. With the prayers of these and kindred spirits was every stone and every brick consecrated to the Lord.

From the well preserved records of subscriptions it appears that some of them were to be paid in installments on the first of January, 1835, 1836, and 1837.

PREPARATIONS.

One book contains the names of more than eighteen hundred subscribers from ninety places, promising a total of $27,000, as follows:—

Abington,	$1,136 50	New Haven,	850 00
Ashby,	180 75	Northampton,	670 00
Ashburnham,	10 00	Northbridge,	32 00
Ashfield,	37 00	North Bridgewater,	202 00
Athol,	158 00	North Wrentham,	5 00
Barre,	60 75	New Hampshire,	4 00
Belchertown,	225 00	Oakham,	66 75
Blandford,	340 00	Palmer,	32 87½
Boston,	6,270 00	Pawtucket,	5 00
Braintree,	73 00	Paxton,	87 00
Brimfield,	37 73	Peru,	977 75
Charlemont,	440 00	Petersham,	65 12½
Chicopee,	35 00	Phillipston,	202 00
Colebrook,	50 00	Plainfield,	757 00
Conway,	1,405 00	Princeton,	52 50
Cummington,	100 00	Rochester,	180 00
Dighton,	5 00	Royalston,	98 75
Enfield,	442 75	Rutland,	34 50
East Abington,	30 00	Sandisfield,	205 00
Easthampton,	1,850 00	Somers,	315 75
East Randolph,	13 25	Southboro',	66 00
Falmouth,	373 75	Southwick,	50 00
Fitchburg,	266 00	South Hadley Canal,	25 00
Florida,	50 00	Spencer,	5 37
Gardner,	30 58	Springfield,	750 00
Goshen,	103 00	Stockbridge,	250 00
Granby,	663 00	Sturbridge,	74 00
Greenwich,	28 75	Suffield,	120 00
Hardwick,	50 00	Templeton,	76 00
Hatfield,	215 00	Thompson,	100 00
Hawley,	55 00	Townsend,	128 54
Heath,	1,200 00	Uxbridge,	50 00
Hinsdale,	275 00	Ware,	208 12½
Holden,	25 00	Warren,	63 00
Hubbardston,	49 50	Westboro',	71 50
Ipswich,	100 00	West Hampton,	200 00
Lenox,	75 00	West Medway,	67 50
Leominster,	25 00	Westminster,	32 75
Longmeadow,	125 00	West Springfield,	350 00
Ludlow,	89 00	Weymouth,	102 00
Lunenburg,	42 50	Whately,	5 00
Middlefield,	1,123 50	Williamsburg,	150 00
Monson,	825 00	Windsor,	200 00
Montagne,	49 00	Worthington,	124 50
New Braintree,	144 75	Wrentham,	800 00

These subscriptions vary from six cents, in three cases, to one thousand dollars from Deacon Safford and another thousand from Samuel Williston. Partly because of the financial depression of 1837 some of these pledges were never redeemed, others only after long delay. Another book has lists of names with pledges of ten, twenty, fifty, or a hundred dollars annually for five years. The larger the sum, the shorter the list.

Of all these documents probably the one Miss Lyon most prized is the paper bearing the autograph signatures of the Ipswich ladies to the first thousand dollars.

THE PARLORS.

CHAPTER V.

FIRST BUILDING ERECTED.

1836—1837.

LEGAL standing was given the enterprise by an act recorded as follows:—

Be it enacted by the Senate and House of Representatives in general court assembled, and by the authority of the same, That William Bowdoin, John Todd, Joseph D. Condit, **David** Choate, and Samuel Williston, their associates and successors, be and are hereby incorporated by the name of the Trustees of Mount Holyoke Female Seminary, in South Hadley, in the County of Hampshire, with the powers and privileges, and subject to the duties and liabilities provided in Chapter forty-four of the Revised Statutes, passed November 4, in the year 1835; and with power to hold real and personal estate not exceeding in value one hundred thousand dollars, to be devoted exclusively to the purposes of education.

HOUSE OF REPRESENTATIVES, February 10, 1836. Passed to be enacted.
JULIUS ROCKWELL, *Speaker.*

IN SENATE, February 10, 1836. Passed to be enacted.
HORACE MANN, *President.*

COUNCIL CHAMBER, 11th of February, 1836. Approved.
EDWARD EVERETT, *Governor.*

The limit of valuation has since been extended to $1,000,000.

March 2nd the five trustees met at South Hadley, accepted the act of incorporation, and added Rev. William Tyler and Rev. Roswell Hawks to the board. April 13th they added Joseph Avery, **of Conway,** and arranged for preparations to build. The next day Mr. Tyler and Miss Lyon drove to Monson, and were joined by Mr. Hawks at the house of Deacon Andrew W. Porter.

Their errand is explained in a letter written by Mrs. Porter after Miss Lyon's death:—

"On answering the door bell on a snowy day in April, 1836, a stranger stood before me, who introduced herself as Miss Lyon. I was prepared to give her a cordial reception, having a high regard for Miss Grant and Miss Lyon as principals of Ipswich Seminary, but what could lead her there that stormy day I could not think. She soon explained. 'You have heard of our contemplated seminary. An act of incorporation has been obtained, the location decided on, and now we need some one to superintend the work of building—one whose business talents have been tested, who has had experience in building, and in whose integrity the community would have confidence; one, too, who would do it without remuneration, for it is all a work of benevolence. Last evening your husband was named to the trustees as one to whom we might apply. Rev. Messrs. Hawks and Tyler were appointed to wait on him, and I was requested to accompany them.' Mr. Porter was in Boston to return that evening. Miss Lyon accepted my invitation to remain till Monday, but both gentlemen had appointments for the Sabbath to meet. It was decided that one of them should return on Monday Miss Lyon retired to her room before Mr. Porter arrived and not a word was said about the seminary till Monday. Ten or twelve years after, she told me those were nights of prayer. 'And the Lord,' said she, 'not only answered my prayer by inclining your husband to engage in the work, but gave me yourself and Mr. Porter as personal friends, and your house as my home. O that first visit and that chamber where the enterprise was commended to God anew, and the question of your husband's acceptance wholly submitted!'"

Deacon Porter was appointed trustee April 19th. May 6th, Mrs. Porter wrote Miss Lyon, "We are deeply interested and see your present emergency. Yet I am sorry to say Mr. Porter does not feel he can take new care now. For three years he has suffered much from his head and I have been constantly desiring him to lessen his business. But I must say that it has been my desire he should engage in this benevolent enterprise and trust the Lord for health. Should he return from his journey with strengthened nerve and should there be assistance he can render, I think he would do it most cheerfully. Our daughter sends love. Since you were here she h s made bread twice, with good success. She means to be qualified for a bread-maker at Mount Holyoke Seminary." But the daughter did not enter; two years later at the age of thirteen, she followed her three brothers to a better school, leaving her parents childless, and yet not childless, for then more than ever they adopted the seminary whose members henceforth they fondly called their daughters.

The partial release from business which Deacon Porter had just secured, Miss Lyon regarded as providential for her cause. He was soon as active in it as if it were his own. Through that season he spent several days each week at South Hadley. The next year, from March to November, nearly every Monday he drove there, twenty-one miles, returning home Saturday. During all this time he left his own extensive business in other hands, provided his own conveyance, entertained himself and horse, and made no charge whatever. For forty years every interest of the seminary continued to be the object of his conscientious care, and constant prayers. But for the co-operation of Deacons Avery, Safford, and Porter, the farmer, the smith, and the manufacturer, it does not appear how the enterprise could have gone forward. Who can doubt that the Lord helped Miss Lyon to find such men and inclined them to enter into her plans when so many wise men could not comprehend them. She loved to trace his hand in providing each of her noble band of helpers. A Holyoke pupil writes: "She wanted us all to know the names of Dana, Choate, Heard, Felt, Packard, and other early friends of whom she seldom spoke without a moistened eye. I had not been a week in the seminary before I had heard of them all. She had so told us of Dr. Humphrey and Dr. Hitchcock that the very mention of their names filled us with reverence. We had heard of the faithfulness of Mr. Tyler and Mr. Bowdoin and were assured that Deacon Safford, Deacon Porter, and Deacon Avery would soon come to see us. And when they came we saw that their hearts were even as Miss Lyon's heart." Another says: "She never forgot to tell us of Deacon Porter's absence on her first visit, and of her two days of seeking to be willing to give up securing his aid, before she could even talk with him; then she would add, 'And now don't you think God has given us in Deacon Porter the very best man His storehouse could furnish?'"

Many, first and last, came into her plans for a time who failed to see all things as she did, and presently withdrew. She often said that the right persons were raised up in every strait and that it was wonderful how they would fall away when there was nothing more for them to do,—doubtless to teach her not to trust in the arm of flesh. But when trustees could not see alike it was a source of solicitude. For a time they were not unanimous in the choice of a site. The first spot talked of was just north of the Eastman place, on a gentle elevation half a mile from the church, which Miss Lyon thought too great a distance. She was not partial to the site chosen May 19th, yet dreaded to have the subject agitated again lest it lead to desertion of the cause by some whom it could ill afford to spare. Two months later, those who desired a more commanding location had become so dissatisfied that the whole board was called together to reconsider. But the vote of May 19th was confirmed July 28th, and soon the turf was broken thirty-five feet from the road, for a building of brick ninety-four feet by fifty, and four stories above the basement, designed to contain public rooms for school and family uses, and private rooms for teachers and eighty pupils; every member of the school to room and board in the same building. It would not accommodate the two hundred planned for, but receipts were less than half the thirty thousand dollars asked for. To delay longer would be to lose a part already pledged, for change and death knew no delay. And additions could be made as funds should be received.

When excavations were nearly done, an apparent defect in the foundation threatened to reopen the question of site. It resulted in removal twenty-five feet farther from the road. One walking now about the grounds in the rear of the building might get no hint of the ravine which limited the removal and for filling which no funds could be spared at that time. Miss Lyon said, "I wish it could have gone much farther back, but this was something I could not control." After the house was done she used to preclude all criticism by pointing out the magnificent views from the upper windows.

Next, a doubt was raised about the bricks procured, causing delay and suspense. But competent judges approved them and once more the work went on. September 21st Miss Lyon wrote George W. Heard, Esq.: "The trustees are now laying the foundations of the first edifice. The corner-stone is to be laid on Monday, October 3rd, at two o'clock, with appropriate religious exercises. Mr. Todd will give the address. We especially desire the members of the original committee to be present. I therefore write to you, and hope Mrs. Heard will be able to come with you." Of October 3rd she wrote Miss Grant: "It was a day of deep interest and tender associations. The stones and brick and mortar speak a language that vibrates through my very soul. How much thought and feeling have I had on this subject. And I have lived to see a body of gentlemen venture to lay the corner-stone of an edifice which will cost about fifteen thousand dollars, for the education of women. Surely the Lord hath remembered our low estate. This will be an era in the cause. It may have to struggle through embarrassments for years, but its influence will be felt. The work will not stop with this institution."

One source of solicitude steadily followed another, yet courage never forsook her. She used to say, "It is one of the nicest of mental operations to distinguish between the very difficult and the impossible." She seemed to have that gift. After the question of site had been settled and re-settled, the foundations relaid, the bricks tested anew, and the walls were slowly rising, the structure fell to the ground. "Now," said the agent, "Miss Lyon will be discouraged," and he dreaded to meet her. But she came to the scene of confusion as cheerful as ever, exclaiming, "How wonderful! no one killed, no one hurt!"

President Heman Humphrey and Professor Edward Hitchcock of Amherst College were added to the trustees October 3rd, and Deacon Daniel Safford of Boston, the following April.

The building must be furnished. Turning again to the ladies of her acquaintance, in December Miss Lyon presents a plan by which the sewing societies of different towns might each provide furniture and bedding for a private room, at a cost of from fifty to sixty dollars, surplus donations being used for public rooms or housekeeping utensils for a family of one hundred. With the plan goes this appeal: "And now, dear madam, would not the ladies of your place consider it a privilege to furnish one of these chambers? Would you not also consider it a privilege to bring the subject before them so fairly that they will do it promptly?" And as if to aid them in presenting the subject, she sets forth the need of the world for thoroughly educated Christian women, and the plan of the seminary to train such women; and calls special attention to the enterprise as a test of the question whether the founding of schools for the education of women shall have place among public benefactions. "In this," she writes, "lies its chief importance. It is like the signing of the Declaration of Independence; the battles were still to be fought, but the question of independence was settled. It is like the fitting out our first band of missionaries; the work of evangelizing the world was still before the church, but the question of acknowledged duty and the mode of meeting it was settled. Let this enterprise be carried through by Christian liberality, and it will no longer be uncertain whether the cause shall stand among the benevolent operations of the day. The work will be before us, but the principle on which it is to be accomplished will be settled. The progress of the enterprise in gaining an acknowledged standing has exceeded the expectations of its warmest friends, although the work of bringing this institution into operation has been longer than was anticipated. Had I a thousand lives, I could sacrifice them all in suffering and hardship for its sake. Did I possess the greatest fortune, I could readily relinquish it all, and become more than poor, if its prosperity should demand it."

In 1835, the general committee with Miss Lyon's cordial concurrence had invited Miss Grant to unite with her in taking charge of the new seminary; her negative answer was not from lack of interest. This was shown in an effort made in Boston, in March, 1837, when Rev. Messrs. Coggswell, Anderson, Blagden, Winslow, Rogers, and Boies, with fifteen or twenty laymen, met by invitation at Deacon Safford's, "to confer with regard to the seminary, and to take measures for advancing its interests." Mrs. Safford, Miss Lyon, and Miss Caldwell were also present by Deacon Safford's wish. He had consulted some of the gentlemen on the point. "They thought there would be no impropriety in admitting us to hear what was said," wrote Miss Lyon to Miss Grant, adding, "Dr. Anderson made some very pertinent remarks and read from your letter to him. I have since borrowed and read it to others, and all are very much interested in it. I could not take a copy of it without your leave, but I want one and will make good use of it." Of that man greatly beloved whose name for half a century was almost a synonym for that of the American Board, we learn from Mrs. Cowles that, "in those days of sneer and obloquy when scarcely one distinguished man gave the plans of Mary Lyon his support, the quiet voice of Dr. Anderson gave no uncertain sound. At a meeting of a few Christian gentlemen in the parlors of Deacon Safford to hear her plans and confer on their feasibility, his unfaltering *yea* gave a decisive turn in favor of her novel enterprise. For years he was her shield."

At the close of the meeting over three thousand dollars was pledged, and a week later the sum had risen to four thousand three hundred and sixty-five dollars. The largest subscriptions were one of one thousand dollars by Deacon Safford, two of five hundred dollars each, and four of two hundred and fifty dollars each. In the letter to Dr. Anderson, so much prized, Miss Grant refers to woman as divinely designed to be the chief educator of the race, influencing not only her scholars and her daughters, but her sons, brothers, and older men, not excepting even her father and his peers; to the unprecedented demand for teachers throughout the land; to the increasing readiness of young women to qualify themselves to teach; and to their lack of suitable opportunity for study; and urges the case on this wise: "I hope the benevolent men of our metropolis will not dismiss this subject without careful examination. The question is not whether the plan of that seminary in all its minutiæ is adapted to their taste, or whether it is as good as their united wisdom could devise; but it is whether they will help build up an institution founded on Christian principles and designed for the education of women. We ask for aid not for the sake of an individual, not for the sake of woman alone, but we ask it for our country, nay more, for the world. My soul kindles as I write, but I am exceeding the limits of my strength, whose failure has been caused chiefly by efforts to sustain an institution without such

ERECTION OF FIRST BUILDING. 69

means as it would be economy for the Christian public to furnish, and such as I think they would long since have gladly afforded if they had understood the subject in all its bearings. I have not a doubt that my labors will be curtailed many years in consequence of my increased burden for the want of what comparatively small funds would have furnished."

Miss Grant's fears proved true. Without endowment, without her former associate, her strength grew less and less, and in April, 1839, she was forced to resign her charge and leave for life the work she loved so well. But her biographer says that grace was given her to look on with thankful heart that the great cause was carried forward, though she herself was held back from the work. It was with her full approval that Miss Lyon secured as pupils the first year at Holyoke, a few from Ipswich Seminary on whom she could rely to give tone to the school; the arguments which she used and which led them to make the change were these: their help was needed and would tell more in the new school, and they might thus share in the responsibility and reward of aiding to found an institution that would do good long after they should rest from their labors. Her two teachers, and her associate principal, Miss Eunice Caldwell, now Mrs. J. P. Cowles of Ipswich, had taught in Ipswich Seminary.

Crowded with work and plans of her own as were the three years before opening Mount Holyoke Seminary, Miss Lyon was not too busy to help others in their need. We have seen her in the summer of 1835 going back to take charge of Ipswich Seminary, that Miss Grant might travel. Even before she had left Ipswich in the autumn of 1834 she was invited to Norton for consultation with Judge Wheaton about the seminary he was founding in memory of his daughter, and in which Miss Lyon was thoroughly interested. The bank note he placed in her hand on leaving, instead of using for stage fare, she gave to the thousand dollars she was then collecting of ladies.

Perplexities of her own did not prevent care for the interests of others. In July, 1836, when dissatisfaction with the site at South Hadley was greatest, after rehearsing various causes of solicitude and delay, she wrote from Norton, "And so I came here to see how the new house comes on." What the new house was will appear in the following quotation from the "Semi-centennial Sketch" of Wheaton Seminary:—

"It is something to remember that before Mount Holyoke Seminary was established, Mary Lyon was busy with plans for the new seminary at Norton. Her own school was of course first in her thoughts, but she took many days, sometimes weeks, to visit Norton and aid in the beginning of the work there. Miss Caldwell, who had promised to go with her to South Hadley, could be spared for two years, and became the first principal of Wheaton Seminary, though not without distrust of herself. But Miss Lyon knew the person she had recommended and was ready to assist in emergencies. There are those who distinctly remember the first assembling of the school on an April morning in 1835. They recall the very tones of the principal's voice as she enters with her cheery 'Good morning, young ladies!' And through their graphic words we see again the brisk figure of Mary Lyon moving among them in the weeks that followed, and hear her quick 'Hasten on, young ladies, you are not aware of the habit of lagging you are forming,' as they passed to recitations. Even her gait is recalled—her business-like manner of moving swiftly forward, which made her seem to stoop as she walked; and her way of bringing a long lead pencil down on her left fore-finger while talking earnestly, her eyes fixed upon distances others could not pierce. 'Do not waste the precious moments,' she was ever saying, and her advice was always supplemented by Miss Caldwell's motto, 'Always in haste, but never in a hurry.' How well the habit of expeditious mental work was inculcated may be inferred from the fact that under Miss Lyon's direction a number of the girls went through Adams' Arithmetic in three weeks.

"The boarding-house was not built till the second year. The plan of bringing pupils into one establishment for a home and for study had not then been often tried and there were fears regarding its success. But Mary Lyon with her clear foresight and strong constructive faculty, in whose mind Mount Holyoke Seminary was already a real edifice, although no stone of its foundations had been laid, insisted that such an establishment must be, and it was. Mrs. Cowles says that Miss Lyon fairly 'talked the boarding-house into being' and took charge of its details, leaving to her the more congenial occupation of managing the school. While the walls of Mount Holyoke Seminary were going up, Miss Lyon's enthusiasm regarding it aroused a strong interest among the Norton girls. They contributed funds for furnishing a room in the new institution. A record at Mount Holyoke Seminary in Miss Lyon's handwriting contains the following entries:—

October, 1837, cash from teachers and pupils in Wheaton Seminary towards furnishing parlor, ... $100.00

November, cash from teachers and pupils of Wheaton Seminary to complete the furnishing of parlor, . . . $135.50

"Those who had no money to give, worked to earn it, one pupil taking the responsibility of keeping the seminary building in order for two weeks, to obtain the two dollars subscribed for that purpose.

"Many Norton pupils followed Miss Caldwell to South Hadley and thus it happens that some prominent names are found on the early lists of both seminaries. This is especially true of some who afterward became missionaries."

After two years as a private school Wheaton Seminary was incorporated in 1837.

In a pamphlet published February, 1837, entitled, "A General View of the Principles and Design of Mount

Holyoke Seminary," Miss Lyon states more fully than in previous circulars, the following points:—

"It designs to fit young women to be educators rather than mere teachers; and to develop the most useful women for any sphere, rather than to supply teachers who shall devote their lives to that profession; and also to establish the principle that the education of the daughters of the church calls as rightfully for the free gifts of the church as does that of her sons." Statistics follow, showing the contrast between Protestant supineness in this respect and the activity of the Papal church in seeking control of coming generations through her girls' schools in this land. The pamphlet closes with an appeal to patriotism, philanthropy, and Christian benevolence for aid to Mount Holyoke Seminary as the representative of the cause of education for woman.

April 12th, the trustees formally appointed Miss Mary Lyon principal and Miss Eunice Caldwell associate principal of Mount Holyoke Seminary. In May a prospectus was issued indicating the academical arrangements, and adding:—

The teachers and pupils will constitute one family, and none will be received to board elsewhere. Except in extraordinary cases none will be received under sixteen years of age. If any must be refused, preference will be given to those who have been teaching.

The school year will comprise four quarters of ten weeks each. The charge for board and tuition will be settled by experiment. In the present fluctuating state of the market the trustees will name a price for one quarter only at a time. For the first quarter they have decided to place board, exclusive of fuel and lights, at thirteen dollars, and tuition at three dollars, making the bill sixteen dollars for ten weeks, to be paid in advance. As far as definite encouragement has been offered, it has been that the regular bills for board and tuition would be from one-third to one-half less than in existing seminaries. It will be seen, on comparison, that the terms stated are not far from one-half charged elsewhere. The expectations of the public will therefore be fully realized even if on experiment it be found that actual cost requires the charge to be somewhat higher hereafter. The domestic department will be in charge of a competent person. All the members of the school will aid to some extent in the domestic labors of the family. The time thus occupied will be so small that it will not retard progress in study, but rather facilitate it by the invigorating influence of a little daily exercise.

ERECTION OF FIRST BUILDING.

The division of labor will be very systematic, giving to each young lady not much variety in a term, but enabling her to perform her part in a proper manner, without solicitude. To each will be assigned that in which she has been well trained at home, and no one will receive instruction in anything with which she is entirely unacquainted. It is no part of our design to teach young ladies domestic work. This branch of education is important, but a literary institution is not the place to gain it. Home is the proper place for this instruction and the mother is the appropriate teacher. Some may inquire, "Why then this arrangement?" We reply, that the family work must be performed—that it is difficult to find hired domestics, and to retain them when they are found—and that young ladies engaged in study suffer much in vigor of body and mind and in their future health, for the want of exercise. The construction of the building and the family arrangements will render it convenient for the members of the school to take part in the domestic department, thus receiving and conferring benefit. Daughters of well-bred families in New England have independence enough to do anything which will promote their best interests, and the best interests of those around them, and for such families this institution is designed.

This feature of the institution will not relieve mothers from giving their daughters a thorough domestic education, but it will rather furnish additional motives to be faithful in this important duty. Is it not a reflection on both mother and daughter, when the daughter cannot perform with skill and cheerfulness any domestic labor which is suitable for her mother?

The plan for the domestic department is an experiment—but one respecting which there are sanguine hopes of success. That the experiment may be a fair one, it is important that the plan should be executed on the principle of entire equality; that the labor should be performed as a gratuitous service; that all should participate; and that none should be received who are entirely unacquainted with domestic work, or who cannot cheerfully co-operate with others in carrying out these arrangements.

In the formation of all the plans of this seminary it is kept in mind that the labors of any teacher are but temporary. Much care will be taken to adopt permanent principles and to mature a system which may outlive those who inaugurate it.

The low price for board and tuition did not accord with the personal views of the trustees. But they knew that Miss Lyon had studied the subject thoroughly and results had so often proved her right, and as one of them said, they had so much proof that the Lord was with her, that they had learned to fear to oppose her plans.

If they were ever as sanguine as she, at that time their hopes were unusually dim. May 11th, Mrs.

Porter wrote to Miss Lyon at Norton: "The committee are becoming discouraged about proceeding with the building, fearing it will not be possible to raise the ten thousand dollars needed before October. Now if you can come it may raise their drooping spirits. I try to encourage Mr. Porter. The Lord can send silver and gold enough for these walls to be built even in these troublous times. It is his cause and I believe he will not suffer it to be hindered. What is the prospect for the furnishing fund?"

In just what way their faith was rekindled by her coming, we have no record; but we learn about the furnishing, by letters from ladies in all parts of the state responding to her appeal in December. Such of those time-stained letters as remain to this day have an interest merely from their existence. They were written, folded, and wafered half a century ago. From the official figures—"6," "10," "12½,"—in one corner telling the postage, and the postmark in another showing the distance they came, we can learn the law of postal rates, while the "paid" on some and the lack of it on most tells from whose purse the postage came. But the value of these old-style sheets to the heart of every daughter of Holyoke lies in the story they tell of courageous efforts, sometimes successful, sometimes not. Some joyfully inclose money already raised, others convey no pledge. One offered to furnish the crockery, another asks if so small a sum as twenty dollars would be accepted, if no more could be gathered. Here is one postmarked "Phillipston, Paid 10." Let us open and read. "The subscriptions to the general fund are not yet all paid. Most of the ladies are dependent for earnings, on the palm leaf hats they braid; with these they can buy materials with which they can make pillows and bedding; but fifty or sixty dollars is more than they can raise; could not some other town unite with them, that both may do what neither can do alone?" Other letters report that sewing societies are finishing comforters and pillows and sewing upon a corner of each the name of their association. Those pillows were sometimes used by pupils unfamiliar with this chapter in Holyoke history to whom those names were a mystery. But here is a letter in Miss Lyon's own hand, to her niece, June 26: "Considerable money will need to be borrowed to finish the building, and the trustees will not feel authorized to hire more. [They had voted that the amount hired should not exceed five thousand dollars.] I hope to secure the furnishing from ladies, but the trying times render it exceedingly difficult. I think not one-third of the amount needed is yet pledged. Everything done for us now is like giving bread to the hungry and water to the thirsty. Sometimes it seems as though I should sink under the burden. How all can be done before the first of November I know not; but in view of what the Lord has done for us we have abundant reason to trust him."

Finding her bodily presence more powerful than her letters she went to many towns, meeting the ladies and inspiring them with new zeal. In July she is again at Norton, where she hears from Mrs. Porter: "The carpenters are nearly done in the two upper stories. It is very desirable that you should come soon on account of the divisions in the basement." Now she must attend to finishing the domestic hall, where no one else can superintend the work; but the furnishing was never out of mind. By September it was plain that new efforts must be made; turning to the expected pupils for aid, she sent a list of articles needed, and asked for a return list of such as each could bring. She expressed the belief that friends if necessary, even at a sacrifice, would cheerfully lend certain articles for a time; the young ladies to whom they were entrusted being responsible for their careful usage and safe return. Before the opening of the school, word was to be sent to each whether all the articles on her list would be needed.

Most of the fall Miss Lyon spent in South Hadley. She was received as a sister at Rev. Mr. Condit's, next door north of the unfinished building. She looked after the drawers, cupboards, closets, shelves, latches, and hinges. One man declared that not a nail was driven she did not see. Another says, "She saw that the church's funds were applied for the convenience and welfare of the church's daughters. When the joiner work was done she made ready for the plasterer; when the plastering was done she made ready for the painter, and when the painter had done she saw to the drying."

Her evenings were filled with correspondence. To the work of finishing and furnishing was added the incalculable care—to use her own words—"of economizing our means and contriving how to do without what we cannot have." Her great inventive powers were called into full play and when one door was shut she always found another open. But no pen can describe the labors or anxieties of those months. Seven weeks before the school was to open she wrote: "When I look through to November 8th, it seems like looking down a precipice which I must descend. I can only avoid looking at the bottom and fix my eye on the nearest stone till I have safely reached it. I try to take the best possible care of my health. I have had more prostrating headache the last few weeks than usual, but on the whole I am very much sustained by a kind Providence."

Friends feared for her health. There was no escaping from her labors, but in all the whirl of work and care she was as punctual at meals, at prayers, and in retiring as she ever required her pupils to be. She regarded all her strength as the Lord's, yet entrusted to her care. We catch the undertone of her spirit in her words to Miss Grant: "Do not cease to pray that no one who contributes money or time or influence, to this undertaking, may ever call aught his own."

November came. Deacon Porter was hastening on the work outside. Deacon and Mrs. Safford were helping indoors. Delays and disappointments continued.

Goods delivered were not always according to contract. The supply of tin-ware was late in coming and then proved inferior to the quality ordered. Longer delay was inconvenient, but sacred funds were not to be wasted, and the lot was promptly returned. For the same reason a lot of bedsteads was about to be sent back, when Deacon Safford interposed. "Let them at least have a trial," said he. Miss Lyon replied, "I am not satisfied to have such beds set up. But if you say so, Deacon Safford, we will keep a few of them." And one of them was set up in Deacon Safford's room. It broke down the first night, and Deacon Safford not only acquiesced in Miss Lyon's judgment, but paid the extra expense occasioned by following his advice.

At the close of the Preparatory Lecture, Friday afternoon, November 3rd, Mrs. Obed Montague, as she tells us, proposed to several ladies to go from church to call on Miss Lyon and see the new seminary before the young ladies assembled. They were cordially shown over the building by Miss Lyon herself. As they were leaving she invited them to a "working bee" the next day to help put the house in order. They accepted the invitation and promised to bring others. Arriving at one o'clock, they met Miss Lyon and a few pupils carrying away brick and shavings left by retreating workmen. Thirty pairs of willing hands were soon at work; some on unfinished bedding, some in cooking for next week's arrivals, and others in arranging rooms. By dark a surprising amount of work was done, and Miss Lyon invited all to stay to tea. They protested, and Deacon Safford said: "Why, Miss Lyon! you know there isn't a tea-kettle in the house—nor any tea." But with Miss Lyon tea meant supper. "Besides," she added, "I want you to test Mrs. Safford's cooking."

"Seldom had Miss Lyon a happier face," says Mrs. Montague, "than at that supper table, so grateful was she for what had been accomplished that short November afternoon."

Her house was getting ready.

CHAPTER VI.

THE OPENING.

1837-8.

IT helps to realize the contrast between the traveling facilities of to-day and fifty years ago, to remember that when the three pioneer railroads of Massachusetts, from Boston to Lowell, Worcester, and Providence, were opened in 1835, there were but one hundred and ten miles of railroad in the state, and but three hundred in the Union, against two thousand in Massachusetts, and one hundred and twenty thousand in the United States, in 1885. While in 1835 it was considered a great success that two daily trains made the trip between Boston and Worcester in two and a half or three hours, frequent express trains now accomplish the distance in eighty minutes.

The first passenger train from Boston to Springfield brought pupils for the seminary; but this was not till October, 1839. One says: "We left Falmouth at four in the morning, Boston at six, dined at the Massasoit House in Springfield, and alighted from the coach in South Hadley [13 miles north of Springfield] about dark." To-day one could go in that time from Boston to Buffalo, either climbing more than one thousand feet upward over the Berkshire hills, or plunging through Hoosac almost two thousand feet below its surface. The first annual catalogue gave the following directions to Holyoke pupils: "A daily stage from Hartford to Brattleboro' and from Brattleboro' to Hartford passes South Hadley. Young ladies from the west, by stopping for the night at Springfield, will find a stage in the morning for South Hadley. From Northampton, they can take an early stage to Hadley and leave there by another stage for South Hadley. Those from the east can pass the night at Springfield, Hadley, or Amherst, and leave by a morning stage for South Hadley. By taking a private conveyance at Belchertown, they can arrive the same evening, and avoid the travel of several miles. A carriage will be sent from South Hadley to meet the stage on its arrival at Belchertown for such young ladies as will write Miss Lyon in season. South Hadley is about six miles [south] from Northampton. Young ladies who arrive at Northampton from the west can obtain a private conveyance from that place in the same way."

The Brattleboro' and Hartford stage was continued until the Connecticut River Railroad was opened to Willimansett, which was announced in the catalogue of 1846. The tenth catalogue states that the railroad conveyance for South Hadley terminates at Smith's Ferry. Though telegraph and railroad were yet to come, the quiet Connecticut valley was not unknown. The "little Nile" had attracted to its banks the first emigrants from eastern Massachusetts, and since the wars of King Philip, King William, and Queen Anne, Deerfield, Hatfield, and Hadley had been historic names.

This valley has many interesting geological features. Its bed was once an arm of the sea extending from Long Island Sound to the northern boundary of Massachusetts. Its bottom and sides were formed by the gneiss rocks on the east, and the mica schist on the west, between which, in some period of disturbance, came up the trap ranges of Holyoke and Tom. The gradual filling up of this estuary by the action of streams, produced immense beds of rock belonging to the formation known as the Triassic or New Red Sandstone. These ancient shore-beds, once covered by the tides, still retain in their strata footprints of the strange animals of that time. Examples of the impressions of the feet of animals, of the stems of plants, and of ripple marks, have been dug out of the rocks upon the seminary premises, and others may still be seen in the bed of the brook. Little did Pliny Moody think, in 1802, when his ploughshare turned up a track in stone on his South Hadley farm, what an interest the scientific world would take in the outcome of that discovery, or that the region would become the most famous locality in the world for fossil foot-marks.

While Miss Lyon was planning the seminary, Prof. Hitchcock was enthusiastically at work in this new field of investigation, the results of which are seen in the valuable ichnological collections of Amherst College and Mount Holyoke Seminary. In the varieties of rock scattered through the valley in the form of drift, and in the numerous rounded hills of the picturesque landscape, the geologist sees evidences of the action of ice during the glacial period. Prospect Hill on the grounds of the seminary, is one of these moraines fashioned by the ice.

South Hadley is in Hampshire county, which, as Dr. Tyler says in his "History of Amherst College," "has long been the banner county of the state in its educational and religious history. Statistics show that it exceeds any other county in the proportion both of its college students and its church members. In 1832, old Hampshire county, with a population of sixty thousand, had one hundred and twenty students in college; which was twice as many as the average of the state. It was then computed that if the whole state sent young men to college in the same proportion, she would have twelve hundred students instead of six hundred, and the United States one hundred thousand instead of six thousand. And whether as cause and effect, or more likely both cause and effect of this, it is now equally distinguished for the number and character of higher educational institutions."

In 1887, within a circuit of eight miles from Northampton there are besides Mount Holyoke Seminary, three colleges, two academies, a seminary for boys, a young ladies' school, and an institute for deaf mutes.

THE OPENING.

There is a tradition that parents in Hadley tearfully implored the blessings of Heaven upon their sons and daughters when they left the old village to settle in the woods south of the mountain. In the "History of Hadley," pp. 395, 396, Mr. Judd says:—

"The Indian war would have prevented the removal of families to the south side of Mount Holyoke earlier than 1725, but there may have been a few settlers in South Hadley then. Their first petition to the General Court was in November, 1727. Twenty-one men represented that they were 'Residents on a designed precinct in Hadley, south of Mount Holyoke,' about eight miles from the place of public worship in Hadley, and the way mountainous and bad. They desired to be a precinct, and to have added a tract of province land on the eastern border four miles long and two miles wide (afterwards named the Crank); the General Court granted their requests, November 28th, provided they had forty families in two years, and should settle a learned, orthodox minister in three years. A second petition of twenty-six persons was presented July, 1728, requesting to be a precinct from Mount Holyoke to Springfield bounds and from Connecticut River west to the equivalent lands east. The petition was granted August 1st, provided they built a meeting-house and settled a minister in three years. In June, 1732, they sent a third petition, requesting that their precinct might be established, though they had not been able to settle a minister in the time set. The Court, July 4th, gave them two years from August 1, 1732, to settle a minister. The first meeting of the South Precinct of Hadley, in the records preserved, was held March 12, 1733; there must have been previous meetings, the record of which is lost."

In 1870 occurred the semi-centennial anniversary of the Sabbath School of South Hadley, when Rev. J. M. Greene, the pastor, drew the following picture:—

"Fifty years ago this town was little known abroad. There was no seminary whose classes every year carried its name around the world; no mills whose paper and cloth bore off millions of labels stamped with its name. The population was only one thousand and forty-seven—now [1870] it is two thousand two hundred. Its valuation was one hundred and fifty thousand dollars—now it is one million five hundred thousand. The occupation of its citizens was almost wholly farming. There were two or three saw-mills and grist-mills, a large oil-mill at the Falls, a mill for dressing cloth at Pearl City, and—we regret to add—two large distilleries and several cider mills. There were three school houses in town, one near the church, one at Falls Woods, and one at South Hadley Falls. [Miss Abby Wright taught a private school for young ladies in South Hadley a number of years, commencing about 1802. It was in good repute.] The Woodbridge school, a boarding school for boys, was somewhat noted from about 1827 to 1834. The merchandise for the town fifty years ago came up the river on flat-bottomed boats and landed at the Falls. The stage driver's horn was heard twice each day in this village, once as he went through from Amherst to Springfield, and once as he returned."

Of the villages in Hampshire county, perhaps none was more quiet than South Hadley fifty years ago. Its one meeting-house expressed a union of sentiment among the people which had great influence in deciding the location of the seminary. The village itself would not be recognized now by one who saw it then, but from the upper windows of the seminary are seen the same views which feasted the eyes that first looked from them.

"Against the sunrising," says a lover of their beauty, "is our own Prospect Hill, which hides all beyond it. Blue, distant hills lie against the southeastern sky, and in the foreground are the pond and the old red mill. Woods and broad fields stretch to the southern horizon, and the brook from the pond winds and twinkles through the nearer meadows until it flows into 'Paradise.' On the west, low, shifting sand hillocks, pine groves, and

MT. HOLYOKE SEMINARY & GROUNDS.
FROM GOODNOW PARK.

cultivated fields hide from our sight the river, beyond which are little foot-hills, the Mount Tom range, and the sunset. Between Mount Tom and Mount Holyoke, the river has made a wide highway through which we look to farther hills, shining in the sunset glow of summer days, as if the celestial city were there. Against the whole northern horizon lies the Holyoke range, one little hill close against another showing where the 'Notch' breaks the mountain barrier between Amherst and South Hadley. Perhaps the loveliest view from the seminary grounds is this Notch seen from a point where trees hide a wider prospect to the east and west."

The same writer continues:—

"Yet many places within South Hadley precincts overlook quite as extensive and beautiful scenery, each having such loveliness, all its own, that in its presence one cries, 'This is finer than all beside.' This river-valley, so famed for its beauty as well as historical associations and geological treasures, has nowhere greater charms than in this immediate vicinity. Who can justly describe the varied loveliness of river and meadows, villages, forests, and hillsides, as seen from Mount Holyoke or Mount Tom? What painter on canvas or by word could picture the splendor of these encircling hills when glowing with autumn's scarlet and gold? What can surpass, in peaceful beauty, the drive through the wide streets of Old Hadley, then down the river-path and under the mountain, just at the sunset hour?

"Many points of interest to the geologist are easily reached from the seminary. The place from which Pliny Moody took the most valuable 'bird tracks' is less than two miles away. 'Titan's Piazza,' a remarkable columnar trap-rock formation, is on the side of Mount Holyoke, 'Titan's Pier' overhanging the river a little farther down. Near it is the 'Pass of Thermopylæ,' a rock fissure, high above and close upon the river, wide enough for a carriage-road.

"Forest and field are rich in the variety, beauty, and rarity of their wild flowers and ferns. Among the hills, beside the river, along country roads and by-ways, and deep in the woods, are countless nooks to linger in and return to, as to a restful friend. To a great number who have been here in the half-century past, the mention of names such as 'Paradise,' 'Chestnut Woods,' 'Bittersweet Lane,' 'Iron Bridge,' recalls 'pure pleasures manifold,' whose influence has never ceased to bless."

The ten acres first bought for the seminary, of Joel Hayes and Peter Allen, lie about forty rods south of the church, on the east side of the main street, and slope to Stony Brook, seventy rods eastward. Huckleberries grew near the road; farther away a blackberry patch was crossed by a rough ravine. April 12, 1837, while the building was in progress, the trustees voted "That the land east of the run be seeded down this spring, and sown with oats, and all west of the run be planted with potatoes, under the direction of the building committee." Ten years later they appointed a committee "to improve the ravine east of the seminary, and to make it suitable for cultivation."

The edifice first built, ninety-four feet by fifty, is that part of the present main building north of the seminary hall. A description written during the first term gives this sketch of the interior:—

"The third and fourth stories have each eighteen private rooms; the second has eight private, one chemical, and three recitation rooms. A hall in each of these stories extends the length of the house and each has two flights of stairs. In the second story a short front hall leads to the upper piazza and another to the upper story of the wood-house.

"Entering the front door the parlors are at the left, and at the right, the seminary hall, fifty feet by forty, extends to the south end of the building. Its white walls are hard finished. The ceiling is painted white, and the floor, like all in the first and second stories, marble-colored. A straw carpet covers the platform; the chairs are maple-colored and the desks before them are of cherry, with green lids. Folding doors open northward into the reading room, which is finished in the same manner. This is twenty feet by twenty and furnished with a well selected variety of periodicals and a few books. Beyond are Miss Lyon's two rooms in the northeast corner. The parlors are each twenty feet square with folding doors between. One is unfurnished for want of funds. In the other, the carpet, cane chairs, and mirror were given by Miss Caldwell and her pupils at Norton. The piano is the gift of Deacon Safford. In the basement is a dining room, fifty feet by thirty; a kitchen with closets and store rooms; separate rooms for kneading, baking, and ironing; and an ample and convenient wash-room. All are so well arranged and neatly fitted up that much of the work, which is all done by the young ladies, is mere recreation. They would consider it a great calamity to be excluded from the kitchen and have their places supplied by domestics.

"The private rooms, eighteen feet by ten, including a lighted closet five or six feet square, are arranged for two occupants each, and provided with a bed, table, drawers, washstand, mirror, chairs, and an open Franklin stove. The stove is in a corner by the window at one end of the room; at the other is the closet and the door into the hall."

A glimpse within other academic walls is given by the *Harvard Register*, in reminiscences of Harvard life fifty years ago by Rev. Dr. Peabody. His reviewer says, in the *Boston Journal*, February 12, 1880:

"In the times of which he writes, the students' rooms were furnished in the plainest manner. Ten dollars would have been a fair auction price for the contents of an average room. No fellow-student of his owned a carpet. There was a second-hand furniture dealer who had a few threadbare carpets which he leased at an extravagant price to wealthy seniors, but not even Southerners, though reputed to be fabulously rich, aspired to that luxury until their senior year. The rooms were heated by an open wood fire, and it was a common practice to have also a cannon ball heated red-hot on very cold days, and placed as a radiator on some extemporized stand. Friction matches were not yet invented, and the evening lamp was lighted with flint, steel, and tinder-box. The price of board in commons was one dollar and seventy-five cents, or as then expressed, 'ten and sixpence.' The kitchen, cooking for about two hundred persons, was the largest culinary establishment of which the New England mind then had any knowledge, and it attracted visitors from the whole surrounding country.

"The student of fifty years ago went to morning prayers at six in summer, and in winter at about half an hour before sunrise. Recitations or lectures preceded as well as followed breakfast, which at the college commons consisted of coffee, hot rolls, and butter. Dinner occurred at half past twelve. There was another recitation in the afternoon, except on Saturday; then evening prayers at six in summer or at twilight in winter, and then came the evening meal, corresponding to the breakfast except that tea took the place of coffee, and bread was substituted for hot rolls. The recitations were mere hearings of lessons without comment or collateral instruction, and the classes were divided into sections so that each student might be called upon at every recitation. Rich provision was made for courses of lectures and by far the largest part of the actual instruction was by means of them.

"The range of study was much less extensive than now. Natural history and chemistry received little attention; French and Spanish were voluntary studies; Italian and German were chosen by a very few; good work was done in the department of philosophy and in the writing of English; but the chief labor and crowning honor of scholarship were in mathematics and the classics."

The day for the seminary to open came before the house was fully ready. The doors were without steps; the windows without blinds; the wood-house was not covered; stoves were not set up; the furniture, delayed by storms, had not all arrived; and much of the bedding pledged had not made its appearance. For a week or ten days longer Deacon and Mrs. Safford and Deacon and Mrs. Porter were as busy as before, Deacon Porter still looking to the out-door concerns and—by the way—giving Miss Lyon lessons in bookkeeping. Deacon Safford worked day after day till long after dark setting up bedsteads and stoves, unpacking and arranging furniture and the like, just as if he were the father of this great family. We quote from other actors in those scenes; says one: "How well I remember November 8, 1837! My father had brought us—four girls—in his own carriage, a three days' ride from Vermont. Leaving home on Saturday, we spent the Sabbath at a hotel in Chester, and thus were able to arrive in good time on Wednesday. How uninviting that plain brick building! The bare walls seemed almost insecure from their narrow height. There were no trees, no fence, and not a blade of grass, but a deep bed of sand lay all around the house. In the absence of front steps we alighted on the back side of the basement at a door opening into the dining room. At one end of the room a group were at work on unfinished comfortables. At the other, tables were spread for hungry travelers. Mrs. Deacon Safford, a royal woman with a lovely face, and Mrs. Deacon Porter, of no less princely gifts, were washing crockery in the great kitchen. Presently Miss Lyon appeared, her face all aglow under the traditional turban, and gave us the welcome of a mother to her daughters. 'Come right up stairs,' she said, 'you have come to help us,' in a voice that had the true home ring. Heart met heart, teacher and pupil were one, and we followed her to the seminary hall. Deacon Safford, with his coat off, was on his knees tacking straw matting on the platform. Looking up with a bright smile he said, 'We are in glorious confusion now, but shall soon be in order.' In a trice our wraps were thrown off and we were on the floor helping. The enthusiasm of the presiding genius of that hall was so contagious that we could not be idle."

Another says, "I helped Elihu Dwight, I think it was, put down the carpet in Miss Lyon's room. New arrivals were constantly announcing,—'Miss Lyon, I have brought the box of bedding from the ladies of our town.' When the boxes were open Miss Lyon set me to marking the bedding and giving it out for the different rooms."

The first writer continues: "At four o'clock, the matting was down, the bell was rung, and Mount Holyoke Seminary opened! Though the sound of the hammer, the plane, and the lathe, was still heard about the house, no other day could be thought of, for this was the one appointed.

"It is said that Napoleon's greatness appeared in the choice of his generals. Miss Lyon had shown rare wisdom in the selection of her aids. She said, 'I have no personal charms to attract pupils, so I have had reference to these in choosing my teachers.' How proud we were of the elegance, grace, and beauty of the brilliant Miss Caldwell, graceful Miss Smith, and charming Miss Hodgman! They all won our love; but Miss Lyon—revered and beloved—how my heart thrills at mention of her name! Yet I little appreciated her in those days of immaturity nearly fifty years ago.

"Supper came at a late hour. There were eighty of us to be disposed of for the night. Not enough mattresses had been unpacked for us all to have beds, even on the floor. We were very weary, but none of the repining, homesick girls of modern boarding schools were there, for we were coping with realities, not fancies, and under the inspiration of a magnetic leader. 'Young ladies,' she said, as she sent some to Mr. Condit's, others to Mr. Smith's, or to Mr. Allen's, 'you will recall these little experiences in the real hardships of the far West.'"

Night provided for, the next problem was breakfast for the eighty. There was plenty of good bread, thanks to Mrs. Safford and that wonderful Rumford oven of Deacon Safford's. But any housekeeper would declare that the few cooking utensils which had arrived, though they had proved surprisingly versatile hitherto, were quite unequal to the demand. We wonder where, how, or when, Miss Lyon slept that night. After eleven she was tapping softly at the door of one of her new daughters to engage her help for the morning meal. The helper slept, for she writes: "Four o'clock seemed to come in a minute. But that breakfast of mashed potato, bread and butter, and cold water—how good it was." Says another: "There was little furniture, but day by day loads arrived and young men from the village very kindly came and helped Deacon Safford to set it in place. Byron and Morgan Smith, Levi Allen, Newton and Elliot Montague, Elihu and John Dwight, helped to put our house in order. I remember when John Dwight was setting up furniture one evening he saw Nancy Everett for the first, but not the last time. How happy we were in making the best of our inconveniences, and how we appreciated every article of furniture as it came.

"We brought our wood and water from the basement and thought it no hardship. We were so glad to have a place in the seminary, for we knew that more than twice our number had been refused for want of room. It was not long before each room had its wood-bin in the corresponding story of the wooden wing, and each story soon had its water cistern filled by a force pump in the basement.

"The South Hadley people were very kind to us. Mr. Hayes showed us to our seats in church as courteously as if we were personal friends whom he delighted to welcome to the house of God. Winter mornings, instead of being shut in by the deep snow, we found the nicest of paths cut for us by somebody in the night."

The building was too unfinished to be formally dedicated before the school opened, and that service was deferred till the 3rd of May following. The manuscript of Mr. Condit's address still exists. A poem was written for the occasion by Sarah K. Browne, author of "The Last Indian," and "The Spartan Mother," then a member of the middle class.

But there never had been an hour when the seminary was not the Lord's. In Miss Lyon's words at the first, "Every brick of this house is consecrated. You must not call this Miss Lyon's school. I regard it so truly a child of Providence that I do not like to have my name made prominent. And you would look upon it as I do if you could see the many gulfs that were to me impassable, bridged over by the divine Hand. Sometimes all seemed to hang upon some slight pivot, without which the whole would have fallen to the ground. I can see a ruling Hand in everything connected with its establishment, and I would have you ever remember that you are studying in an institution built by the hand of the Lord."

It was not the building alone, nor the seminary in general, which she consecrated. "How modestly did she tell us," writes a pupil, "that as the founder of the seminary, she had made a covenant with God in behalf of every pupil who should ever enter it. She trusted that God would preserve it to millennial days and use it for his glory; she had made us over to God and all who should follow us. This covenant on her part involved obligations on ours. We could not lightly disown it. Tenderly she besought us not to live henceforth unto ourselves, but unto Him who died for us and rose again." Miss Fiske adds, "Miss Lyon continued, each succeeding year of her life, to impress this responsibility upon her pupils, and perhaps no one ever left her to be happy in living for herself."

CHAPTER VII.

THEORY PUT TO TEST.

1838.

IT was Miss Lyon's theory that a seminary could be founded and conducted on the principle of benevolence.

The building and its furniture were free gifts. Her teachers wrought with her in the spirit of the true minister or missionary, being supported, not compensated. And every pupil had come with the understanding that she was to contribute to the carrying on of this plan of benevolence by sharing in the care of the household, not as a servile labor, but as a benevolent service. Objectors, who had been sure that money would not be given, or if it were, that such teachers could not be found, still believed that this third feature would fail; and many warm friends feared the same.

It is noteworthy that though the economy of this feature was at first a reason for its adoption, Miss Lyon had discovered others so much stronger that she hardly alludes to that in the circular of February, 1837. She saw that it would relieve from dependence on private families for board, a relief essential to permanent prosperity; that it would relieve also from dependence on hired domestics; and that the interest of the young ladies in home duties would be preserved, while the daily exercise would promote health and happiness. Before the school opened, so thoroughly had she become convinced that this plan was desirable, independently of its pecuniary advantages, that on one of her visits to Ipswich she sought to convince Miss Grant that it would be wise to adopt the same in Ipswich Seminary. And now this part of the theory was to be tested. Here were four-score people to be provided three times a day with food cooked by their own hands, and a great house to be kept in order, all without infringing upon school work. It is no small thing for a matron, even with well trained servants, to keep in order so large a boarding-house. It is yet more difficult lovingly to lead so many girl students to do it. To find a lady to whom the literary interests of the seminary could be in a degree committed was comparatively easy, but Miss Lyon could rely on no one else to organize the domestic department, and at first she gave this her chief attention.

To the skill of a Napoleon in finding generals, she added the tact of an Elizabeth in discovering what each one could do and putting her in the right place. But then the very best one to aid in preparing dinner might be reciting in the geometry class at eleven o'clock. What was to be done? It was easier to change the hour of a recitation than to fill that place on the dinner circle. But to alter the recitation-hour might interfere with the engagements of some one else and require another change. Never had Miss Lyon more frequent use for her wondrous powers of invention. When for the twentieth time the literary and domestic departments interfered, for the twentieth time she readjusted her time table, and as cheerfully as at the first. She had often said at Ipswich that she could suggest plans by the score leaving Miss Grant to select which she chose. So, at Mount Holyoke her resources never failed, and in every exigency the right order of exercises appeared in due time.

The combined care of school and family, that first winter, demanded from sixteen to eighteen hours of the twenty-four. Her celerity was wonderful, and yet she could scarcely answer the calls for her counsel that came from all parts of the building. The smallest details of household cares were faithfully provided for.

Not only had everything a place, but she knew when it was in place and how to keep it there.

No wonder that, as she sometimes said, her head seemed full of bread, tin-dippers, and clothes-pins. But she saw "Holiness to the Lord" written on everything in the building and hence her minute care that nothing should be misused, not a window-sill be defaced, nor a "dust of flour" wasted. Surviving pupils recall the precept, "Never burn what a bird would open its bill to get." Through the domestic department they learned many a life-long lesson of economy, order, and faithfulness in that which is least.

To students bread is emphatically the staff of life. Considering the quantity needed, the season of the year, and the manner in which the work was done, no practical housekeeper would be surprised if the first attempts were unsuccessful. "We have the best of flour," said Miss Lyon, "we can have good bread, and we must have it!" Not one in the house had ever before seen a Rumford oven. Selecting the most reliable pupils, she took the lead herself. Her writing desk was carried to the basement, and by snatches she conducted her large correspondence while watching the processes in the baking room. This she did till she had herself learned and taught her helpers all the mysteries of bread making. Her roommate at that time relates: "One day the bread was poor. As she told me of this new disappointment, Miss Lyon leaned her head on my shoulder and wept. Her girls must lack for another day the light, sweet bread they ought to have, and which she had tried so hard to give them. Soon she wiped her tears, passed to the inner room and shut the door. The next day's result led us reverently to say, 'The secret of the Lord is with them that fear him.'" To Miss Lyon's view whatever was necessary to the health and comfort of her family was as vitally connected with the cause of Christ as direct labor for the salvation of souls. Morning after morning in the dark and cold, she rose to watch the rising of the bread with an eye as single to the glory of God as she rose to pray. With the simplicity of a child, she mingled prayer for souls with prayer for the work of her hands, expecting and receiving a sure answer.

One of the bread-makers tells us: "In the evening Miss Lyon convened us in the baking room and proposed a new plan. It was carefully adopted. In the early morning we hastened to the basement. The light sponge was all ready for us. How glad Miss Lyon was! Eyes, voice, hands, feet, testified. She was so glad she could not stand still. 'See,' she said, 'our difficult problem is solved! No more poor bread—no more interruptions in lessons and recitations! Now our arrangements will run like clock-work!'"

In this connection, the same pupil gives another picture of Miss Lyon: "In compliance with her request—'Come to my room and let me know'—I have entered to find her with her Bible, so absorbed, half in study, half in prayer, that knock or voice failed to rouse, till a hand rested on her shoulder. I think it was that same intense earnestness welling up from the heart, filling face and voice with something irresistible, that has left its impress upon so many of her pupils."

By December, Miss Lyon wrote: "In their domestic work the young ladies are all that I could wish. I should not have supposed that in three weeks they could go forward with so much system."

Near the end of the first year she wrote to Rev. Theron Baldwin, principal of Monticello Seminary, Illinois;—

> On the whole the success of our institution in every department is greater than I anticipated. I am more and more interested in the enterprise as a means of developing certain principles of education for woman, especially the importance and feasibility of introducing system. Our experiment is entirely in favor of limiting the age. We have definite requirements for entrance: and though we thought it expedient to admit pupils even for one quarter the first half-year, and for a term the second half, we now venture to require all to stay a year except in extraordinary cases. We can receive but about ninety. We have already had two hundred applications for next year. We have a fine class of students in scholarship and character.

But your inquiries refer more particularly to that appendage of our plan—the domestic department. It succeeds beyond my most sanguine expectations. Its advantages are far greater than I looked for. The difficulties and immense labor of organization were also much greater. It was far easier to find all my teachers than one qualified for the head of this department. In this respect I had been quite mistaken. Though the lady employed would do well for many places, we soon found she could not meet the demands of this. In all such cases you know there is but one course to take. We must do the work ourselves. When we have an interest in planning we can sometimes make up in zeal what we lack in skill. The failure of my superintendent was a great disappointment, but it was not without its advantages. Every part of the plan of organization is the result of personal observation. For weeks I was engaged many hours daily, contriving about furniture and cooking utensils, and planning the division of labor, and for times and places so that everything could be done in season and in order without interference with studies or recitations. I had several points to gain. One was a high standard in the manner of doing the work; and another was that every part of it should be in favor with the young ladies. For three or four months I did not leave the seminary even for a half day. I then said that I considered the family organized and that I wished to go to Boston for a rest of some weeks and to see whether the wheels I had been so long in adjusting could run without my aid. On my return I found everything in order, and there has not been a time since when I could not be absent three months without sensible injury to that department. I need not go to the basement once a month now, though I like daily to pass from room to room to see how delightfully all goes on. " I trust you will excuse this egotistical description. Its object is to give the facts as they are. But to be more definite: the work is done by circles, each having a leader; They are formed for one term; every young lady is responsible to be on the spot at the appointed time, and the leader is responsible that the work is done well. One circle washes and keeps in order the crockery; a second washes and rubs knives a third has the care of the glass and silver a fourth of setting tables; a fifth of sweeping public rooms (private rooms are cared for by the occupants); a sixth of making bread; a seventh of preparing pastry; an eighth of baking bread and pastry; the ninth, tenth, and eleventh prepare the meals.

The young ladies like to keep our house as neat as the neatest. They never object to washing floors—all are painted—twice a week or every day if needful, nor to rubbing knives every meal. How unlike common domestics! I have been a boarder for more than twenty years, but never had everything done for me so well as now.

Our circumstances are so favorable that our case is scarcely a test for other institutions. In the first place we have no pupils under sixteen. Nearly all are from good New England families and are conscientious and efficient, generally well taught and well trained in housework. Secondly, we have no domestics. At first I thought we might need one or two, but now think we are better off without them. If anything is not quite so pleasant, the query never rises whether it is more suitable that servants should do it. No one feels she is doing a work from which she could be relieved by paying money. We hire a man to take care of our garden, saw wood, and do various things for our comfort. Thirdly, our family is so large that by a proper division of the work all can be done if each gives to it about an hour a day.

Among the incidental advantages of the system may be named the union of family interest it secures, the social intercourse, and the healthful exercise—more healthful than if taken for its own sake merely. The older and more studious are generally more inclined to be negligent in this respect, especially in the winter. This plan gives each an hour of exercise daily and at the same hour. No time is wasted in debating whether it shall be taken or omitted. It proves no hindrance, but a help to mental activity. Social freedom and vivacity add to its benefit. One young lady said she was somewhat homesick at first, but the first washing day was an effectual cure. The home feeling is strengthened, and each member becomes identified with the family. We have but one interest instead of the three separate interests of most boarding-houses,—that of the head of the house, that of the boarders, and that of domestics. The unfavorable effect of these separate interests I regard as one of the greatest objections to sending daughters to boarding-schools. It endangers the simplicity, kindness, and mutual confidence which have been so tenderly fostered at home, and tends to develop artfulness, selfishness, and distrust.

I have no confidence in mere theory in education. Our plans are a combination drawn from experience and practical observation. Thus far we have been enabled to accomplish on every point all that we have encouraged the public to expect.

In calling this department an "appendage of the plan" Miss Lyon alludes to the fact that though she saw more and more reasons in its favor, she did not regard it as an essential feature, but one which could be modified or abandoned as circumstances might require. Though she succeeded in reducing it to admirable order, and made wheel move within wheel without friction, she often said: "This department is too complicated and requires too much care to be continued were it not for its great advantages. If dollars and cents alone were concerned we would drop it at once. Had I known how complicated its working must be, perhaps I should never have undertaken it; but a kind Providence hid many of its difficulties and I see so much in it for the comfort of the household and favorable to each member individually that I am willing to take all this care. Young ladies at school with all the helps and comforts which they should have, naturally incline more to being ministered unto than to minister to others. To counteract this there is needed the special cultivation of an unselfish spirit, while opportunities for its cultivation are comparatively few. To bring every such opportunity to bear on the character is a leading object in our arrangements. We would furnish an example of a Christian family. In their varied and mutual domestic duties the young ladies daily find many occasions for exercising a generous and self-denying spirit, whose influence will be felt through life. The system also helps to cultivate a sense of obligation and gratitude. Home life is little less than a continuous conferring and receiving of favors. Domestic happiness depends on their being conferred with a willing heart and received with suitable tokens of gratitude. These traits go hand in hand. The formation of a grateful disposition is specially important in a lady's education. Parents should seek to give their daughters privileges, and especially the means of an education, in a manner suited to lead them to realize that they are favors for which gratitude is due."

One to whom the position of domestic superintendent was offered, replied, "Perhaps some of my friends might consider it dishonorable." But Miss Lyon saw only honor in mutual ministries for comfort, and no dishonor in waiting upon one's self. Such ministry and self-service from a worthy motive ennobles any labor. That she did not object to things repulsive to a worldly or self-seeking spirit, appears from her playful words: "This feature also serves as a sieve, holding back the indolent, the fastidious, and the feeble—of whom we never could make much—and giving us the finest of the wheat, the energetic, the benevolent, and those whose early training has been favorable to usefulness."

It is not strange that Miss Lyon's lofty aims were misunderstood, nor that her methods should fail to be appreciated in advance; and no clearer proof can be given that the system is still above the common comprehension, than the frequency with which the seminary is referred to as a "manual labor school." Even in 1885 the historian of another seminary in New England remarks, "It has never been thought best to introduce the industrial element as at Mount Holyoke." It would be equally appropriate to call any family which employs no servants a "manual-labor family"; or to ascribe to it on that ground the introduction of an industry. But a false impression sometimes seems ineradicable; and there is yet occasion, even in Massachusetts, to repeat Miss Lyon's words: "It is no part of the design of the seminary to teach housework; that would make it far too complicated and expensive. However important this part of a woman's education, a literary institution is not the place to acquire it." "But only the other day," writes a recent graduate, "a clergyman told me in good faith that cooking and general housekeeping *had always been taught* at the seminary. He would hardly believe me when I assured him it was not so. I asked him how it could be, when for perhaps half the year the same young lady only dusted recitation rooms an hour a day or wiped dishes, or set tables; how many different things could she learn in four years? 'O, but they change every once in a while, and so in the whole course they learn everything,' was the reply."

The writer adds: "However, I believe we did learn to take pretty good care of our own rooms. I remember how mortified I was when one of the teachers blacked her fingers on our dusty window sash."

The success of the feature the first year was enough to mark it as a stroke of genius, yet that success scarcely lessened opposition. Says Dr. Hitchcock in the memoir of Miss Lyon, "'How long must it be tried to satisfy you,' I asked of a friend who thought it must fail. 'Five years,' he replied; and although when the five years had ended the success was still complete, he was no more satisfied than at the first. Others said that when the novelty was over it would be unpopular and must be abandoned. When they found that five or ten years only gave it greater perfection,—when on visiting the seminary they saw how admirably affairs were managed and how inviting was the food,—then they predicted that as soon as Miss Lyon should be removed this feature must be given up. Yet more than two years have passed since that event and never was the arrangement more satisfactory than now. I know not what other period of time will be fixed upon for it to come to an end, unless it be the close of the present century. In that case present unbelievers at least will be spared the mortification of confessing that Miss Lyon's judgment in this matter was better than their own."

The details of the system are constantly being modified as circumstances require, but the principles Miss Lyon adopted remain unchanged. The better they are understood, the more they are prized.

Holyoke students of to-day, exempt from the heavier kinds of work, enjoying the elevator, steam for heating and for culinary purposes, with a matron to superintend the cooking and a man to take all care of the oven and its use,—can only partially appreciate the inconveniences of earlier days or the noble spirit of their predecessors.

That first year brought together a heroic band; nearly all were professing Christians—young women of lofty aims, and steady devotion to Christ. Most were over twenty years of age, and some had suspended their studies for two or three years that they might finish them at Holyoke. Four entered the senior and thirty-four the middle class. Three were assistant pupils. Never were gathered eighty more willing hearts or nimbler hands. Their zeal for the new seminary was scarcely inferior to Miss Lyon's. Their ambition was to vie with one another in self-denying labors for its prosperity. They loved it the more for the sacrifices they made, for the toils they shared with their leader. They counted it an honor to aid in carrying out her plans. Catching her spirit, the love of Christ constrained them no less when employed in household cares than in worship. Scattered in many lands, they have been almost without exception servants of Christ, asking neither thanks nor praise from man, seeking no reward but the consciousness of entering into the work and sufferings of their divine Master. Some have reached the goal and received their crown; others are still serving or suffering.

We have seen that Miss Lyon had placed the charge for board and tuition, against the advice of trustees, at sixty-four dollars for the year. Provisions were high that year. Never did financier more carefully husband resources. Her biographer says:—

"At the close of the year when her accounts showed the trustees that the income had more than met the outgoes, their incredulity vanished. They saw that she understood business. For the next sixteen years the annual charge was sixty dollars. Gladly would the directors of many a corporation pay thousands for such financial skill as she exercised, almost to her own cost.

"Let it not be supposed that Miss Lyon's labors that first year were limited to domestic and financial interests. Besides giving systematic religious instruction, she matured a course of study, watched the recitations, directed individual students in the selection of studies, criticised compositions, instructed the middle class in chemistry—performing with them a course of experiments, and taught several other branches. For the first time in her life she taught Whately's Logic, and entered into it with as much eagerness and relish as she had plunged into Virgil in the days of her youth."

The year closed Thursday, August 23rd. On Monday and Tuesday there were public examinations in the seminary hall. Wednesday, while part of the school attended the commencement exercises in Amherst, others prepared for the forty guests of the next day. The closing examinations were on Thursday forenoon and the graduating exercises in the afternoon, with an address by Rev. Dr. Hawes of Hartford. He was a friend of Mr. Condit and was entertained in his home. A friend of the family says: "After attending the examinations for a time, Dr. Hawes came to Mr. Condit's and asked to be left undisturbed in his room, explaining to Mrs. Condit that he left home without a very high opinion of the 'manual labor school,' but that he had come from the seminary hall to give himself—without regard to meals or late hours—to preparing an address more worthy of the occasion, for it would never do to present anything he had brought with him."

In Miss Lyon's account of the week to Miss Grant we find these items: "The question of going to the meeting-house Thursday afternoon came up once or twice and was settled in the negative, as I felt a great reluctance to it. Wednesday evening I found that the trustees and others were becoming decided that it was best to go. I thought it the most modest to acquiesce. The certificates were given at the close of the services, but no other exercise differed from a common public meeting. It did not appear unsuitable as I thought it would, and I was glad I consented. Our certificates were signed by Miss Caldwell and myself and countersigned by Mr. Condit, the secretary of the board. They were presented by Mr. Condit in his own neat and elegant manner."

The diploma is in English. At the top is a vignette from a design by Mrs. Dr. Hitchcock, illustrative of the words beneath, "That our daughters may be as cornerstones, polished after the similitude of a palace."—Ps. cxliv. 12. The seal of the seminary, bearing a similar device, is attached by a ribbon to the parchment.

The following is from Miss Caldwell's pen: "The trustees, the orator of the day, the teachers, the senior class, and the school, walked to the church in procession, the school clad in white, with heads uncovered, and shaded by parasols. The side pews and galleries were already crowded when Miss Lyon led her beautiful troop in quiet dignity to the seats reserved for them. It was an hour in her life never to be forgotten. The battle had been fought, the victory was hers. In all that year she had never found an hour to spend in astonishment at her success, but now, when circumstances forced the view upon her, wonder, gratitude, and praise filled her heart. Her great soul was surcharged with joy; smiles and tears strove for the mastery on her radiant face. For an hour she resigned herself to the emotions of the occasion and gave way to a joy with which no one could intermeddle."

THE ART GALLERY—NORTH ALCOVE.
Lyman Williston Hall.

CHAPTER VIII.

METHODS ADOPTED.

WE have seen that Miss Lyon designed the seminary for the training of young women for the greatest usefulness. In her view the best educated Christ-like woman was the most useful woman. Looking beyond their own wishes or the purpose of parents, she received each pupil as from Christ himself to be molded into his likeness and trained for his service. She regarded herself and associates not as mere teachers nor yet as educators only, but as moral architects.

Her work in character-building was based on the principle of self-help in all lines, physical, mental, and moral. She desired pupils of sufficient years and maturity to have some degree of mental power and self-reliance, yet not beyond the age when habits are easily formed. To their new and more favorable circumstances she adapted methods which she had tested at Ipswich and elsewhere, shaping every arrangement of both school and family with reference to her one aim.

She constantly had regard for the health of her pupils. That sound health was not so general among our country-women fifty years ago as many seem to believe, is shown in the following sentences from the anniversary address of Rev. Dr. Anderson in 1839:—

"There is cause for much alarm in respect to this matter. A physician declared a few years since that not more than one adult woman in ten in the circle of his observation, enjoyed complete health. In no other civilized country is there such deficiency of health among the more educated women, such a proportion of them

'Too weak to bear
The insupportable fatigue of thought,'

as in our own land. The fault is less in the school than in the family; less with teachers than with parents."

Miss Lyon's eye was on future mothers and teachers also. She chose a healthful location; secured an unfailing supply of pure water; adopted the best methods of the time for lighting, ventilating, and warming every room; and provided sufficient and wholesome food, purchasing only the best of its kind, and allowing no inferior standard of cooking; indeed, in every known way she steadily planned to secure the best sanitary conditions. By requiring regularity in meals and in hours for rising and retiring; a daily hour of work in the house and another of exercise in the open air; regular calisthenic practice; and clothing suitable for the climate; by instruction in the laws of health and the consequences of their violation; by counting exposure of health as no less faulty than neglect of study; by line upon line and precept upon precept she strove to lead each pupil to aim to have a sound mind in a sound body.

The studies prescribed were not those only which develop the imagination and refine the taste, but primarily such as "strengthen the practical faculties, mature the understanding, and lay a firm basis for character." She insisted on thoroughness in preparation and progress. Few were allowed to take more than two studies at a time. Weekly reviews prepared for the general review required in every study before it was left for another, and a certain standard must be attained as the condition of advancement. Recitation by topic four days of the week led to easier use of the pen in the essay work of the fifth. Appeals were made to the highest motives only, and no prizes were offered, no rivalry stimulated. Cultivation of class feeling was avoided. Family interests were nobler than class distinctions. The family was the ideal unit, not the class. To be once received as a daughter was to be ever after one of the sisterhood and dear to the heart of Alma Mater. Miss Lyon coveted earnestly the best gifts for all, but she wished those and only those to finish the course, whose influence would bless the world. She believed that the higher the standard adopted, mental as well as moral, the more valuable would be the attainments, even of those who could not graduate. She knew that many could stay but one year. To them and to all she strove not so much to impart knowledge as the key of knowledge, and aimed to lay such foundations that each should be able to go on in study whether her school days should prove few or many. Taught to place mental power above mere acquisition, they learned to regard education as an unending process, not a finished attainment, and to consider its continuation a duty. The same principle was applied to the seminary itself; not to keep pace with the progress of the age would be to fail of the highest usefulness; accordingly the course of study was extended, and increased advantages were offered as fast as public opinion and pecuniary means allowed. In every way mental culture was regarded only as a means of more effective moral power.

The Bible was pre-eminently the Book of the house, and instruction in its truths was as systematic and thorough as in literature and science. The Scripture lesson was the first to be recited in the week. Not only was more time given to it in regular lessons than to any other study, but its precepts were in constant use. It was read morning and evening in the presence of all. Three mornings in the week Miss Lyon occupied from fifteen to thirty minutes with the assembled school in illustrating and enforcing the teaching of some selection from the Old or the New Testament. She delighted to unfold the great principles of God's government in his works and word, in providence and in grace. She used to say she should not have known how to guide her large family were it not for the history of God's dealings with his ancient people. "If we would learn of God let us read that history. If we would know ourselves, we shall find our hearts well portrayed there. More knowledge of human nature is to be derived from its study than from any other source."

Her range of subjects included the evangelical doctrines, the ten commandments in their order, the sermon on the mount, the book of Proverbs in course, the connection between the law and the gospel, and such specific topics as Consecration, Responsibility, Doing Good, Economy, Regulation of Desires, Cheerfulness, Health, Use of Time, Forgetfulness, etc. Though she taught no formal system of theology, her instructions were always based on some doctrinal truth. In the business and the familiar talks of the afternoon exercise, Bible principles were scarcely less prominent, for they were constantly applied to the great variety of practical subjects discussed.

Once a week Miss Lyon gathered about her the church members of her charge for instruction in their more specific duties. The others she met on Sabbath evenings, and led each to see from the Bible in her hand that if she should fail of fulfilling the highest end of her being, it would be not because she had broken the law of God, but because having broken it she did not accept offered grace. With exceeding vividness and almost irresistible tenderness the claims and the invitations of Christ were set before them, with the responsibility of acceptance or refusal. When her watchful eye saw that the word of God was proving quick and powerful in any of her audience she would publicly invite to her room at a given hour those who desired more personal instruction. In that consecrated place, each going alone, perhaps to find her friend or roommate there on the same errand, many hearts were opened to rejoice in the truth as it is in

Jesus. These were gathered into a class for special nurture.

How Miss Lyon felt about this part of her work appears in expressions like the following: "None but God knows how the responsibility of giving religious instruction weighs on my heart. Sometimes in preparation, my soul sinks with trembling solicitude which finds no relief but in God. When I am through I can only pour out my heart in prayer that the Spirit may carry home the truth." "Everything I do is such a privilege. It is so blessed, too, to depend hourly for light and strength and for success on our Heavenly Father through Jesus Christ our Redeemer." "I want to ask you, my dear friend, to pray for me in a very special manner about one thing. It is for divine guidance in religious instruction. Pray that I may have hid in my own heart all that I attempt to say. Pray that in every jot and tittle, I may speak the words of truth—that which God sees to be truth. Pray that hearts may receive it in sincerity and faith. Pray that in all these seasons God may be glorified."

The use of the Bible was combined with prayer. God was inquired of to do those things which his promises pledge him to do for waiting souls. The seminary was born of prayer. Its principal was a woman of prayer. As she left her closet her face was often radiant, and the experience that shone through her morning or evening talks revealed a communion with God too intimate to be described. She taught her pupils to pray, and showed them how freely they were bidden to bring all their grief for sin or from any cause, all their joys and all their needs to their heavenly Father. She taught them to intercede for others and for the world. She inspired them to concentrate their thoughts in study that they might have power to control them in prayer. They learned to love the time and the place which she secured to each for private prayer, morning and evening. She never asked how that half hour was spent, but simply whether it had been free from intrusion. Morning and evening incense rose also from the family altar. Every week and many times besides, the teachers met to pray for themselves and pupils, singly and collectively. Sabbath evening, while Miss Lyon met the unconverted, her teachers gathered the rest in praying circles. In the Thursday meeting all these circles were united. Timid hearts and tremulous voices gained courage and strength in the smaller gatherings to help in the larger.

In the fifth year of the seminary the students began daily prayer meetings in different parts of the house, during the fifteen minutes recess between the two study hours of the evening. Though a subject was assigned for each day of the week there was always opportunity for special requests. In the following year these gatherings were put in charge of the section teachers. Each invited to her parlor the members of her section. Thus originated the daily "recess-meetings," which have never been discontinued.

The Sabbath was the most important day of the week. All attended public worship both morning and afternoon, according to the New England custom of that time. The hallowed influences of the day were not allowed to be dissipated by visits or calls, either made or received. On the same principle, letter-writing was discouraged.

From the first there were annually two special days of prayer, and sometimes more. On the first Monday of January and the day of prayer for colleges—then the last Thursday of February—school exercises were suspended to give every one the opportunity to join thousands in Israel in fasting and prayer for students and for the world. Their observance, whether by prayer or by prayer and fasting, was voluntary with each one, but so universally was the opportunity embraced that those were days of more than Sabbath stillness. They had a solemnity of their own, and were always anticipated and found to be days of special blessing.

Miss Lyon constantly besought for the seminary the intercessions of its friends. Her parting word was often like the whispered entreaty to Miss Fiske, "You will pray for us, will you not—all the way to Persia." And Miss Fiske knew, as all who have been in the family know, that the absent daughters are remembered unceasingly in their school home. Let a page from the seminary journal illustrate this. "At teachers' meeting Miss Lyon proposed that we mention the names of those once here who are now missionaries either at home or abroad. So each of us named one or more till the names of all were repeated. We then united in two prayers for them, Miss Lyon leading in one." At another meeting all who had ever taught in the seminary were named in the same way. Under a third date the record runs: "When Miss Lyon asked whom we wished to present for prayer, two, not Christians, were named. We knelt and prayed for them, then two more, and so on till twelve had been mentioned." Again: "To-night after repeating their names, we prayed for all in school who were without hope in Christ." Though names might be unheard in the larger meetings, the same definiteness in prayer was everywhere encouraged.

Specific prayer and specific labor went hand in hand. No one was lost sight of in caring for the whole. Partly for its direct help to each, and partly for the guide it afforded to intelligent labor for all, an opportunity, early in the school year, was given the servants of Christ to give their names as his followers. In order to do the utmost for each, in reference to health, habits, and both mental and moral improvement, the family was divided into sections of fifteen or twenty, and each section made the special charge of a teacher. This teacher felt a peculiar responsibility for the spiritual condition of each in her section. Yet Miss Lyon still kept every one in her eye and heart. In the assignment of rooms and roommates she weighed the power of mutual influence and desired such associations only as were most helpful to all. Individual choice had free expression and was superseded only for weightier reasons. Unwise intimacies were everywhere kindly discouraged.

Her own personal influence so permeated the family that every member felt its power, and was assured not only of a warm place in her heart, but that she took a tender interest in her welfare. With scarcely an exception they were glad to have her know all that was going on, and delighted to consult her. If, as sometimes occurred, they wanted to do something which she thought not best, she would often so unfold the principle involved, and carry her audience with her, that they would vote against the course they had previously determined to take.

"There was something in her first meeting with her pupils that cannot be fully described," writes one; "we forgot the teacher in a mother's welcome with a magnetic sympathy welling up from her great heart, that filled face and voice and manner. She was so glad that you could not help being glad with her." Another says, "I almost feel even now the imprint of the greeting received forty-five years ago as my sister and I timidly presented ourselves at the front door where Miss Lyon was standing. 'What name?' she asked. We answered. 'Ah! Miss T— from D—,' and stooping, she kissed us both very tenderly. Then we were adopted. I seem still to feel that kiss a holy thing; and so I regard every association with her, especially every word from her to me individually, as a sacred trust for which I must render account."

Miss Lyon's teachers were one with her in aim and spirit, heartily seconding all her efforts. The trustees had allowed her to choose her assistants and she chose a band of helpmates after her own heart. She never asked of what denomination they were, but she assured herself that the love of Christ constrained them and that their zeal was according to knowledge. Those of the first year were all from Ipswich Seminary, after that they were her own alumnæ. The three graduates of the first year became teachers the second. She sympathized with them and leaned on them as on older daughters. She was as ready to receive as to give suggestions and made them feel that she was grateful for their help. Her manner of referring to them before the school showed that they had her confidence and that she expected their wishes would be gladly complied with. To the teachers she said, "Never speak lightly of a pupil." "Speak of each as if she were your sister." "Avoid every unnecessary exposure of her faults;"—and they did so.—" Don't feel that all is going wrong because some are irrepressible. These lively girls—rightly directed—do the best work." " This girl is a little inclined to be wild. She is motherless. Won't you look after her? her mother was very dear to me."

However it had been in her own home and however warmly adopted in her new home, one could not feel in the seminary that she was "the only child." She was a daughter, indeed, but only one of many, and was reminded of the great principle that in a community each one must give up somewhat of natural rights and consult the general good. She saw at a glance that what would not be best for a hundred others to do or to have in the same circumstances, could not reasonably be done or asked for by one. This ever recurring lesson was used to cultivate the habit of placing the general good above personal preference or convenience. The self-subordination involved was part of the training Miss Lyon deemed essential to self-government. "Be perfect in all the requirements here," she used to say, "and you will have power to control yourself anywhere." Every seminary regulation was shown to be included in the first or second great command. To disregard it was to disregard the law of love to God or love to others. Every one could see for herself its justice and its propriety; and that its observance was her duty, whether it was a rule of the school or not. By the test questions, Is it right? Is it in accordance with the law of love? she was taught that all things were to be done as to the Lord and not as to teachers. Thus the conscience, enlightened by the word of God, was educated to act habitually in all matters of daily life both small and great.

Conscience and a sense of honor were cultivated by the trust reposed. Excluding espionage, Miss Lyon held each pupil responsible for her own observance of seminary regulations and for keeping her own daily or weekly account of success or failure, and trusted her truthfulness in reporting it. She did not expect of all the same attainments, nor equal progress. Her standard in scholarship and conduct was given in the precept "Do the best that *you* can do, *to-day*." But she did expect each one to do right, and assumed that she had no other intention. If she saw reason to fear otherwise she found private opportunity to ask, "Are you doing the best you can?" "Do you not wish to improve?" adroitly preventing self-committal on the wrong side, and pointing out the way of self-help. She knew how to concentrate and combine moral influences. Wisely and warily guarding against the abuse of freedom or of confidence, she sought to have the law so hidden in the heart that no direct exercise of authority would be required, and thus to govern from within rather than from without; and with rare exceptions she succeeded. Often those who had grieved her most became her warmest friends. When for her own good or that of the rest it became necessary to send one away, it was done with the same tenderness with which she had been received, and she went forth knowing that she was followed, not with gossip, but with prayer.

An early graduate writes: "One divine truth, illustrated by Miss Lyon's methods, is the supreme value of love in every effort to do good." It led her to assume that every one had a benevolent spirit. It was an understood premise in every appeal that "to know the need would prompt the deed." Free from selfishness herself she never seemed to suspect it in others.

Were helpers wanted anywhere, the question "How many would like to do" this or that, came as an opportunity to those waiting for one. Was self-denial involved? The end to be gained was shown to be so desirable that sacrifice for its sake appeared a privilege; candid souls said, "If somebody must do it, why not I?"

Miss Lyon never cared to secure the immediate end so much as self-training in benevolent action. So in gifts of charity, she valued less the amount than intelligent and prayerful interest in the call for it, and the habit of liberality with the means at disposal whether large or small; and therefore kept her pupils informed concerning the progress of the Lord's work at home and abroad and gave them the opportunity to share in its support. She thought it essential to the cultivation of right principles that students while spending for themselves should also spend for the Lord, not excusing themselves under the plea that the personal outlay was to fit them for the Lord's service, lest they form the habit of feeling that their offerings must first serve themselves.

Every year, and each time in a new way, she gave a series of morning talks upon benevolence and the Bible standard of giving. Inculcating the spirit of David, who would not offer to the Lord that which cost him nothing, she taught how to look for ways of economy and self-denial, and not only enabled those to find them who thought they had nothing to give, but made giving a matter not of impulse, but of principle; and believing that "practice makes more lasting impression than any amount of instruction without it," she provided stated opportunities for offering gifts to the Lord through well approved channels, and herself set a worthy example. To her associates she said: "We can train benevolent workers only by being benevolent ourselves. The Levites had no portion among the tribes; the Lord was their inheritance; but out of their living they gave their tithes to the Lord. Let us live in the same spirit."

Her views of personal duty are thus expressed in "The Missionary Offering": "I felt that in the sight of God, my duty in my own little sphere and with my own feeble ability was more to me than the duty of all the world besides. Could I call thousands into the treasury of the Lord, it might not be so important a duty for me as to give from my own purse that last farthing which God requires. Could I so plead in behalf of the perishing heathen that all our missionary concerts should be filled with hearts prostrate together before God, it might not be so important a duty for me as to carry my own feeble petition to the throne of mercy, and there in the name of our blessed Redeemer plead the promises with an earnestness which cannot be denied."

Thus impressed herself she laid upon her pupils a sense of personal responsibility in every department of life, especially toward every soul under their influence, and reminded them that theirs was no ordinary responsibility, for the seminary was sacred to the service of the Lord; its founders expected and had a right to expect that it would be a fountain of good to the world, and that the cause of Christ would be advanced by means of it; and therefore no one had a right to avail herself of its opportunities to carry out selfish plans of her own. By receiving advantages so much greater than the price they paid for them, by the unwearied care of their teachers, and by all other unbought benefits, they were under obligations which they could discharge only by doing similar work for others. It was made a matter of daily practice. She depended on the older pupils as on older daughters to help her lead the younger. It was easy to show, especially in the domestic line, that the faithfulness of each member in the duty assigned her was essential to the good of the whole family. The same principle was as plainly shown to apply in their divinely appointed relations to the entire human family. They were sent forth from the seminary, not to sit down in idleness, but for earnest work. "Go," she said, "where no one else will go, not seeking the praise of man, but the favor which comes from God only." "If work needs to be done, and no one wants to do it, that is the work for you. Much of the work of the world, if done at all, must be done for love—not for pecuniary returns. Never decide hastily that you cannot do because you have not physical or mental strength. Every one has something to do for Christ and each is responsible for doing her part, and in the best way in her power. Other things being equal, you are under more obligation because of your opportunities here. Privilege and responsibility go hand in hand." "It is a serious thing to live, to have responsibility not only for your own life, but for your conscious and unconscious influence. No act and no word can be known to be without future consequences."

Lessons upon responsibility were not the only ones for which the domestic department furnished a fertile field for illustration and an ample one for practice. The prompt, expeditious, conscientious, and every way faithful helper about the house—and no other—was the prompt, expeditious, conscientious, and every way faithful student. Hand and head and heart were each trained to strengthen each, but the manual and mental were only for the sake of the moral. In the words of Miss Jessup: "The whole system is an arrangement for getting and applying moral power." And yet it is a progressive system. So broad are its underlying principles and so natural its adaptations that it can easily be kept abreast of the times without loss of its characteristic features.

Miss Lyon often said that she was only laying foundations. She expected others to carry on the work, and rejoiced that they would be able to do far greater things than were possible to her. She would never admit that the prosperity of the seminary depended upon her or any successor. She felt that an infinite Hand had taken hold of hers, and led her on. "The seminary," she said, "is his, built by his direction. I have no more expectation **that it will die than that I shall cease to exist in eternity."** "I doubt not these walls **will stand** to do his work in the millennium." **Not the same** pile of brick and mortar and **not necessarily the** same methods. It was **not Miss** Lyon's **school** that she wished to establish, but **a** school to furnish the best education. Whatever was needed for this end was to be adopted. It would grieve her to have her methods or arrangements made the standard for **her** successors merely because they were hers. She cautioned her pupils against following her methods **too closely**, especially in teaching children or those who were weak in moral principles. **For** their safe use she emphasized the need of sufficient self-control to give **reason** and conscience the **ascendency over impulse and inclination.**

A graduate of 1843, after **long** experience **in teaching**, writes: " **In** my earlier work I was much helped by **the models I had** at the seminary. My later methods were **changed in** many respects, but the great principles of conscientious thoroughness and subservience of the intellectual and esthetic to the moral and religious have **always** lain **at the foundation of** my teaching theories. In the light of my later experience, **the** moral and religious culture of the seminary seems to me a wonderful embodiment of heavenly wisdom. In that department indeed, **methods** are relatively of **less value. The** spirit is everything, **and that** is a fresh gift with each generation of teachers, with each **entering class."**

CHAPTER IX.

METHODS ILLUSTRATED BY QUOTATIONS FROM PUPILS.

A HOLYOKE pupil was once called to undertake a work of public importance, which would cost her great earthly sacrifice. Nature struggled; conscience pressed the claim. The scale was turned for duty, sacrifice, and God, by recalling not what Miss Lyon said, but what she was.

Another says: "It was Miss Lyon's faith and life that gave such indescribable power to her words. The truth she uttered had in her a living illustration so lovely and so majestic as to attract and awe the mind. She seemed to inscribe 'Holiness to the Lord' upon every household duty; on our studies and even on our amusements. I thought then, that perhaps in the course of my experience, I should find some one equally disinterested, and as devotedly pious; but in these forty years I have never seen her equal; and in her memoirs she is not described as so wholly devoted to Christ as she really was. Indeed, I do not think any pen can do justice to her unvarying devotion to her Saviour; eternity alone will reveal its power. With so spiritual a leader the atmosphere of the seminary was religious. Few could resist the influence of such daily life."

A clearer knowledge of Miss Lyon will be gained from the following quotations from her pupils, who repeatedly affirm that her character was more powerful than her words, indeed, that the power of her words was in her character.

"In the first year of the seminary, when commencing housekeeping under many inconveniences, we were on very familiar terms with our leader. I remember how patient she was with us about the turban [which she began to wear after loss of hair from a fever nine years before]. I feel guilty and ashamed to this day every time I see her dear face in that last picture. Why did we not let her wear her turban always, and the becoming arrangement of smooth dark-chestnut hair? I don't remember who began it, but it was the fashion then for ladies of her age to wear caps of lace, tied under the chin, with wide full ruffles around the face, and smooth plain bands of hair underneath, and we wanted her to be in the fashion. So we made up a purse, and bought a beautiful cap of silk blonde; and also wash blonde, and thread lace edging, of which we made others. Then one or two of us carried them to her and asked if she would please to wear them. 'I thought,' she said, 'I should always wear a turban. But I will do almost anything to please my daughters.' She wore caps always after that."

The next tells "who began it."

"But one thing annoyed us. It was that huge turban. My roommate and I, two of the youngest pupils, were clear starching in the laundry one day, when Miss Lyon came in and began to talk of dress, and its influence. 'O Miss Lyon,' I impulsively exclaimed, 'if you only would wear caps!' 'Do you think,' said she, 'that my influence would be better? But then the time! I can do up a turban while the bell is ringing for dinner.' 'You need not mind that,' said we, 'when you have ninety daughters who would gladly do up caps for you.' 'O well,' said she, 'if my ninety daughters prefer to have me wear caps, I am very willing to do so.' The next week a beautifully becoming cap came to her as a Thanksgiving present and the obnoxious turban was seen no more."

"In the affectionate welcome of our first meeting, and afterward through all our intercourse, Miss Lyon impressed me as a woman of remarkable tenderness combined with great force of character. She made me feel at home with her at once. I never was afraid to

Affectionately yours Mary Lyon

speak to her any more than to my mother, and yet I always felt toward her the greatest reverence. Noticing that with her fresh complexion, caps were more becoming trimmed with white ribbon, I said to her one day, 'I wish you would always wear white ribbons, you look so much better in them.' She smiled and replied, 'It would be too extravagant, as they get soiled sooner.' I playfully remarked that if that was all, we would take up a contribution, and buy them."

"She enlisted us in making our home attractive. In the spring vacation of the second year we were requested to bring flower seeds or shrubbery and were promised that the grounds should be prepared and the yard fenced by the time the term began. This was done, and I returned with a trunk filled with roots and seeds, and planted the first dahlias that grew there. Before the summer term closed, the front yard was gay with blossoms."

"She liked to have us free with her. She would rather have our loving confidence than our worship. Suggestions about household arrangements she received with thanks and acted upon them."

"At first the rules seemed many, but I discovered that when trying to do right I rarely violated any or thought them irksome. I once said to Miss Lyon, 'I don't see the necessity of so many.' She replied that at first the law of love was enough, but as pupils increased and were less mature, she had to be more explicit. They came one by one, sometimes at the suggestion of pupils themselves. 'Well,' I admitted, 'they are all reasonable but one. I can see there should be a rule for retiring and for rising, but why one against rising before five o'clock? What would be the harm if one did?' Then she spoke of some who would take time from sleep for study, and of one who wanted to get through in two years and used to retire at ten and rise at midnight to resume her studies, and so injured her health. I saw then the wisdom of that rule."

"My sister had been in the seminary a year, and Miss Lyon had been at my home. I had gained a most exalted idea of her piety. On account of the full house and incomplete arrangements, on my first night in the seminary I was sent to sleep with Miss Lyon. To my surprise she said nothing to me on religious subjects, but told me various ways of going to sleep if inclined to be wakeful, and I think I told her of my mother's way. Even in so simple a matter, I think now how wise she was. The next day several of us were standing in the new seminary hall, scarcely knowing what to do with ourselves, when Miss Lyon entered in her quick way. How matronly she looked with her full figure, her cap, her rosy face, and her eyes beaming with delight and affection as she looked around on her daughters.

"She came to ask—would we like to help make the carpet for this floor? It lay there ready cut. She told us how she had succeeded beyond her hopes in getting the addition to the building, and then came this carpet, an unexpected gift. It was like a mother talking to her girls, of whose co-operation she was perfectly sure. With one accord we went to our rooms for thimbles and needles, and with good heart worked upon the carpet. I was the youngest in school and being undisciplined was continually though unintentionally giving her trouble. Many times she sent for me. I generally came from those private talks with tears in my eyes, and the feeling—'I will never, never disappoint Miss Lyon again.' But the same thing happened time after time. In the long vacation I seemed to gather strength, and in the second year in the midst of a talk to the school she alluded to the difficulties of the new scholars, and said if they would persevere, things would grow easy; great changes for the better often took place, and the second year would develop character. I scarcely know the words she used, but her eyes rested on me with such a look that I thought were I in heaven and she should look at me so, I could ask for nothing more. This perhaps shows how youthful I was, but I still love to remember that look. She often had it—a look of all embracing love. I did not understand it then; I know

now that nothing but the overflowing of a great heart could produce it. When we were helping Miss Fiske in her outfit for Persia, Miss Lyon said to me, 'I want you to be fitted for as good and as great a work as to go on a mission.' This seed thought has never left me, but has inspired the aim to do my work as faithfully as if it were the work of a missionary. A few years afterward I visited the seminary with a friend. Miss Lyon received us like a mother, and with all her cares, came to our room at night to see if we had all the things we needed. This unusual kindliness was not confined to her pupils. Dr. Brooks said that she once requested him to conciliate some who, not through any fault of hers, were prejudiced against the seminary, and added, that if they could have heard the kind feelings she expressed toward them, they could not have retained their prejudice."

"Through the more than forty years of an extremely busy life since graduation I have felt Miss Lyon's personal influence; often in conscious recollection, oftener in the force of some principle which she inculcated. For instance: 'There is a best way to do everything.' Her comment was, 'Study to learn the best way and then practice it. It pays to do the smallest things in the best way. Why, there is a best way to fold an apron!' And even now I seldom fold an apron but her words come back to me."

"One great principle which she inculcated was the subordination of individual inclinations to the welfare of the community. It made it easy to observe rules. Minor inconveniences were more than balanced by the habit of exalting the general good above our private preferences. But beyond this was that noblest of all lessons, self-sacrifice. By precept, by example, and by occasions for their practice, the Christian duty and the value of self-sacrifice were impressed upon us. The remark, 'We must always consider the good of the whole,' was repeated so often that it became a proverb. Yet we knew that in the yearnings of that great heart not a single one of us was ever lost sight of."

"Many have thought Miss Lyon must have been too much occupied with great things to take interest in minutiæ, but while she was devoting six hours a day to hearing recitations, conducting the morning devotional exercise, and the general business of the afternoon, besides attending to the individual wants of one hundred and thirty pupils, she found time to understand the spiritual needs of every one of her charge. How lovingly and tenderly she led us to Christ! And if any one was in trouble she went directly to Miss Lyon, and her great motherly heart entered into our trials just as if she had nothing else to think of. For the motherless and unfortunate she had particular sympathy."

"In the discouragement of new experiences I was at my work one day in the domestic hall when Miss Lyon's beaming face appeared at the door. She probably saw at a glance how matters stood, and almost before I knew it she was at my side. That such a busy woman should stop to comfort a lonely, homesick school girl gave me a deep impression—which grew deeper the longer I knew her—of her kindly sympathy and interest in individuals."

"One incident comes to mind. I regretted it much at the time, but have since been glad that it occurred. In my junior year I was one of a quartette bound together in school-girl friendship. We were not intentionally naughty, but Miss Lyon called us her wayward children. Her keen insight into character saw something from which she inferred that our mutual influence for another year would not be for our good; and at the close of school she told us in her own peculiar way, that she did not think it would be best for more than two of us to return, and that she would write us her decision. Those words of reproof are among my dearest memories, and the look of yearning love, the hand placed caressingly on my head, will never be forgotten. We went to our homes. The letters followed. Her kind heart could not bear to discriminate, and she bade us all return, but gave us her views about our continued intimacy. That letter is among my choicest treasures. Two of us returned to pass one more year under her guidance, and graduate the next. The others declined the implied conditions. In the light of mature experience, I, for one, have never doubted the wisdom of her advice."

"*December* 14, 1848.—Miss Lyon's theme this afternoon was mutual influence. She had noticed a few whose influence over each other was not the most desirable, and said she should probably speak to them individually. She illustrated from chemistry. Some of the rankest poisons are made by the union of the most valuable and inoffensive elements, as, for instance, oxygen and nitrogen. So some young ladies, who were harmless oxygen and nitrogen by themselves, if brought together would make nicotine or strychnine."

"As a teacher she filled her classes with her own enthusiasm. If a young teacher had not learned how to awaken interest in a study, let Miss Lyon appear before the class, and straightway there was infused into it such delight in that subject, that it would become the topic of conversation at all times. Her thoroughness was equally marked, and her way of making us grasp general features. In history, for example, she did not aim so much to teach details as to show us how to pass along the great highway of time, and note the chief events, their connection and their dates, and by repeated reviews, so to fix them in mind that they would never escape. That excellent topic system and those frequent reviews, so stamped on my memory what I learned, that it is as fresh in mind now as forty-five years ago. Enthusiastic in every study, she took special delight in natural religion, and when teaching the sciences, never omitted an opportunity for impressing on our minds the power, wisdom, and goodness of God as displayed in his works. All instruction was made tributary to divine teachings."

In this connection, the following from the pen of Dr. Hitchcock, then president of Amherst College, is pertinent: "Next to religious interests Miss Lyon was distinguished for the thoroughness of her instructions. She delighted in developing principles and hence led her pupils with keen relish even into the dry details of grammar. She had great versatility, and in whatever department of literature or science she was engaged, especially in the latter, an observer would suppose that to be her favorite pursuit. It seemed to make little difference whether it were physical, mental, or theological science, for all these she taught almost equally well. But perhaps the two subjects in which she most excelled, were chemistry and Butler's Analogy, subjects usually thought to demand talents of quite a different order. In almost all her schools she lectured on chemistry and performed the experiments with success. Those who attended the public examinations at the seminary, must have noted the thoroughness of her instructions. On one of these occasions, we overheard one college president ask another on the platform, 'How is it that these young ladies recite in Butler so much better than our senior classes?' 'I do not know,' was the reply, 'unless it be that they have been better taught.'"

"Her knowledge of human nature led her to adopt 'notes of criticism' as a way to remedy many little defects. They served also to bring to light any wrong-doing and were very efficient in her hands in training us in the little amenities of life. Criticisms of table manners —never of persons—were included. Her peculiar manner of reading them was often enough, but if the offense was repeated, her ludicrous—not bitter—sarcasm never failed to effect a cure. Would that all young ladies had teachers gifted with like power to drive away slang and unladylike manners."

"*December* 14, 1847.—At our family meeting this afternoon Miss Lyon's subject was the duty of preserving health. The theme was suggested by one of the notes of criticism, 'wearing thin shoes and cotton hose.' She spoke of the inclemency of New England winters, and of the necessity of additional clothing. She said that women now are less able to bear fatigue or exposure than those of twenty years ago, but that all members of this school are expected to have maturity of character and moral principle enough to do right, without a formal command. If they have not, they should by all means go to a school for younger persons, where they could receive the peculiar care needed by little girls. As she pursued the subject her vivacity increased and she said: 'There are two things, young ladies, that we expressly say you must not do here. One is, that you must not violate the fire laws (alluding to regulations of the family in regard to fire); the other is, that you must not kill yourselves. If you will persist in killing yourselves by reckless exposure, we think by all means it would be better for you to go home and die in the arms of your dear mothers.' She said such exposures were a direct violation of two commands, ' Thou shalt not kill,' and ' Thou shalt not steal'; for a violation of the first robbed the world of the good they ought to accomplish in it."

"There is not a day of my life in which I do not recall those afternoon exercises. My increasing experience of life increases my admiration both of Miss Lyon's ability to select topics, and her power to make permanent impressions. She entered into those exercises with so much heart that one might think that there could be no thought left for the evening meeting. But she could turn from one to the other with perfect ease because always done in the service of the same Master."

"If any domestic duty was neglected she had such a pleasant way of making one thoroughly ashamed, that the fault would never be repeated. I distinctly remember one lesson on order. Having finished my work for the day, I had just reached my room in the fourth story one morning, when this message was sent to me: 'Miss Lyon wishes to see you below.' On meeting her she very quietly said, 'Miss ——, you have left your dish-towel on the table.' It was put in place and I never forgot it again."

"Miss Lyon succeeded in correcting tardy habits. Who can forget how we stood by our chairs at table, till the last loiterer was in her place! and as we heard the distant footsteps drawing nearer, who was not glad that she was not the one upon whom all eyes were fixed. If Miss Lyon's discerning eye saw in the delinquent, one who had repeatedly been late, her words were something like these: 'One minute lost by ninety persons, makes ninety minutes gone forever.'"

"She had great power in communicating her own convictions. She knew how to set in motion a current which made individual opposition as powerless as chaff before the whirlwind. With consummate skill, yet apparently with ease, she moved her pupils as she chose. She certainly secured general co-operation in her plans for the school. On one occasion a leader in the senior class was summoned home by a sudden death in her family. The sympathy of the whole class was awakened. On the day she left, they decided to attend devotional exercises in black dresses and trimmings. The fifteen minutes preceding were spent by all the classes in practicing calisthenics. While they were thus engaged word of the plan to appear in mourning reached Miss Lyon. Messengers were instantly dispatched to the teachers in charge, to request the members of the senior class to come at once to her room. In a few words she convinced them that there were serious objections to their plan,—that it was unwise; a few minutes were given them to change their attire, the bell was a little delayed, when all appeared in dress that did not distract and so disturb the exercises. She felt that sobriety of manner was better there than mourning display."

"Possibly her cap was slightly displaced sometimes. So might have been Joshua's helmet when he was taking Jericho. And though, as has been said, 'her great soul was unconscious of her body save as it was crowded for want of room,' she was not indifferent to personal appearance.

"Her instruction in regard to dress was: 'It should be such as to please God, not laying aside taste, for is he not much more pleased when his children look well than otherwise? I have no idea Christ was negligent of his dress. His garment was one counted worthy of casting lots upon. Taste should be made a matter of practical education. Self-respect is promoted by proper attention to dress.' 'There should be a due correspondence between what we spend on ourselves, and what we give to God. By this rule we should not buy a twenty-five dollar shawl and make a twenty-five cent contribution.'"

"Miss Lyon was careful about establishing precedents. I went to her once with a question of right. I hoped I had become a Christian. My mother was sick at home, and waiting anxiously to hear from me. I said to Miss Lyon, 'Shall I write on the Sabbath?' She replied, 'Can you not write Monday morning before the mail goes?' 'No; I have a lesson in logic at eight o'clock.' After a moment she said, 'I will excuse you from your lesson.' It was a little matter, but it settled the question of letter-writing on the Sabbath from then till now."

"The characteristic of Mount Holyoke teaching seems to me to be its training for usefulness. 'Live to do good'—was the motto. 'Make personal sacrifices for this end.' 'I want you all to teach,' said Miss Lyon, 'if it is only your little brothers and sisters. I do not think a lady is educated till she has had some experience in teaching children. It is a valuable preparation for influence. In no other way can the principles of the human mind and heart be so well learned. If you do not succeed at first, teach till you do succeed. Prepare thoroughly for every exercise and for every recitation, but study the minds and hearts of your children more than any book. I do not expect many of you to give your lives to teaching; but she who can control the minds of the young happily and rightly, is all the better prepared for any sphere. But never expect to govern others till you govern yourself.' 'Do not say, "I would like to take a few music scholars, or assist in an academy." Take hold where no one else will.' 'And never teach immortal minds for money. If your object is money making, be milliners or dress-makers; teaching is a sacred employment.' 'Think of its influence even in this life. The teaching of the children decides the destiny of the nation. It is not necessary to meddle with politics. Educate women and men will be educated. Let all women understand the great doctrine of seeking the greatest good, of loving their neighbors as themselves, let them indoctrinate their sons in these fundamental truths, and we shall have wise legislators. All our statesmen, rulers, ministers, and missionaries must come under the molding hand of mothers and teachers.' 'It is important to be prepared to be good mothers; you can then easily become good teachers, and will in any case be good members of society. It will no longer appear a matter for which no previous training is needed.'"

"No part of our education was for selfish ends. To have and not to impart was to be a miser. The intellect was not placed above the heart. The development of all that is beautiful in moral and spiritual life was put before mere intellectual acquisition. Is not this different from the modern college idea?"

"In the Butler class she spoke one day of the text, 'The secret of the Lord is with them that fear him.' Girls have many secrets with their intimates. The grandeur of being admitted to intimacy with the King of kings was made overpowering. In that class also she referred one day to our highest aims and hopes, saying, 'Some of you desire to be first-class teachers, others would be cherished wives; but, young ladies, if you are God's children, and his glory fails to be your highest aim, in that other ambition you will be disappointed.' The experience of nearly fifty years verifies her words."

"In the third year of the seminary, Miss Lyon watched the progress of a slow fever, expecting the pupil's recovery, but saw at last that she must die. Wearied and worn she left her for a time, and asked me to call her if a change occurred, saying, 'When Jesus comes for one of my dear children, I want to go with her just as far as I can. I do not expect to pass over with our dear friend, but perhaps the Lord will give me a word to comfort her, and it may be that as I see heaven open I shall get a new view of its blessedness to give to those who remain.' It was midnight when I called her. The eye of the dying one kindled brightly at her approach. Kneeling by her side and taking her hand, she said in her gentlest manner, 'Jesus has come for you, Adaline; you will not be afraid, will you? He will carry you safely over. You have nothing to do but to look directly to him. You will suffer only a little longer.' Then she prayed in words as simple as those she had spoken; and in the smile we saw on speechless lips we read the comfort given. In all the years since that November night, I have thanked God for the lesson I learned then—to speak only of Jesus to those for whom he is waiting."

Referring to the same event, another writes: "An incident comes to mind which illustrates the power of a few words from our beloved teacher. During the first few weeks of the year when the entering class was being examined, and when the home longings were fresh in our hearts, one of our number had died. Her soul was at peace with God, but her disease was particularly distressing, and produced great depression. I, for one, felt it terrible to be alone, and shuddered to pass the room so lately the scene of suffering. When we assembled in the hall, Miss Lyon's subject was the privileges of the Christian in life and in death; and she made it appear a blessed privilege to be a child of God in this life however short. When she spoke of the trying illness, the sudden summons, and the desolate home circle, death seemed full of dread. But when with radiant face she told of the release from sin and sorrow, the safety from temptation, the full assurance of a Saviour's love, and perfect bliss throughout eternity,—our spirits seemed all ready to enter upon the joys depicted. My dread was gone. That room of death seemed but the ante-room where the freed spirit had paused a moment before entering in to be with Jesus and behold his glory."

"One night I went from her meeting with the new converts, to help her in arranging for breakfast. Both were thinking of those great themes, and little was said. Stopping in the midst of our work, her face lighted with heavenly radiance, she exclaimed, 'Glorious salvation! Isn't it,—glorious salvation!' Her shining face and rapturous tones are present as though it were yesterday instead of thirty-seven years ago."

"But at morning devotions the intense convictions which were her greatest power, were most apparent. It was there that she impressed upon us the power of littles, little habits, little sins, little indulgences, transient thoughts,—the thousand and one things included in development of character. 'Young ladies,' said she, 'remember if you are God's children, your life, every hour of it, must be a life of care lest you sin against him.' That word 'care' coming from her lips—how heavy it sounded, full-freighted with responsibility!"

"The bright sunlight streams into the seminary hall on an October morning in 1845. Two hundred girls look expectant toward the platform with its row of teachers, as Miss Lyon walks rapidly in and takes the central seat. A snow-white cap keeps in order those auburn locks. The simple print dress is relieved by a plain linen collar. But who thinks of her dress? She has just been in communion with God and the light lingers in her face. With eyes full of a mother's tenderness she looks upon us a moment before her hearty 'Good morning, young ladies!' which we rise to receive. Her reading of the most familiar hymn gives it new meaning and beauty. She wishes us all—like the great family in heaven—to praise God in singing. When she reads to us it always seems to me there are treasures in her Bible that were left out of mine. Her scripture readings are illuminated as well as illustrated. Searching with her either the Old Testament or the New, we learn that Christ is found on every page of both. After prayer we go to our rooms with a new sense of a personal Saviour walking by our side.

"Later we see a model New England housekeeper seated on the little platform of the domestic hall, surrounded by the new scholars. She quickly ascertains the aptitudes of each, and assigns accordingly. Divining the capacity of those who know nothing of work, she makes each feel assured of her ability to do whatever is laid upon her. To me there is more power in this, than in her wonderful exegesis.

"Now we meet in the seminary hall for the general exercise of the afternoon. Several months have wrought improvement in all, but a subtle consciousness that not all is as it should be with some, possesses Miss Lyon. Her habitually sunny face is serious now. Her theme is wrong-doing. Eager, upturned faces give inspiration to her words. While she talks she notes here a drooping eyelid,—there a face flushes under her gaze. The secrets of those hearts will soon be known to her, for a call to her parlor, over whose portal might be written, 'leave concealment behind,' and her kind, faithful exhortation will unlock the inner chambers of thought."

"Her energetic way of saying the most common things constrained us to attend. Her tones, always natural, varied with her thought. Many a pupil could imitate her in saying, 'You won't do so again, will you, dear?' But in the reading of a hymn or in speaking of eternal themes, her voice—low and reverent—had a pathos which could not be imitated, nor forgotten. In the lighter topics of her afternoon talks, it took a wider range. In a gleeful mood, quite natural with her, her tones would fairly dance, and her merry speech and laugh would send waves of laughter through the hall."

"It was a privilege to stay on a week with Miss Lyon after the close of the year. These occasions brought her quite familiarly near. I sometimes contrasted the light-hearted manner of those days, with her solemn bearing just before the meeting with the impenitent. She had asked me to be responsible that the little table for her Bible should be in place. It seemed like getting near the Holy of Holies, to enter her room at these times—for her soulful countenance reflected the result of the preparation hour. Sometimes a word of benediction fell upon me; but usually she was then too weighed down by a sense of responsibility to allow many words."

"Her morning addresses were the crown of all her other efforts; they were very tender and powerful presentations of Divine truth. Her look as she sat without a gesture, her hand laid devoutly on her well-worn Bible, was not produced by the inspiration of mighty thoughts simply, but by a profound realization of spiritual things. We felt that she came to us from sitting at the feet of the great Teacher, with a fresh baptism of his spirit, and a fresh message for us from him. She seemed at times to dwell in sight of the glories to be revealed. What she spoke of was reality to her, and hence her words had indescribable power."

"She said little by way of entreaty. Sometimes she would lift the veil and give us a glance into the Holy of Holies. When the soul was enraptured with its glories, she would turn and say effectively: 'But there will be no vacant seat there. If any one chooses to turn away from Christ, to separate from her Christian friends, her absence will not be felt in that happy throng. Heaven will be full without her.' She would carry the soul on into the unending future, and describe its ever-increasing susceptibilities for joy or pain, as if she had herself been through it all. It was not the words nor the manner nor the thoughts, but the whole effect, which was wonderful. It was the conception she gave us of the truth. With a sense of present real-

THE SEMINARY HALL.

ity, we felt that a thousand years hence we should remember as but yesterday, sitting on those seats and accepting or refusing these offers of mercy. Thus she worked her way down to the depths of the soul and planted seeds to germinate and yield fruit in after life."

"Our note books convey little more than hints of her power, for that lay in her earnest spirituality, and the notes seem like dry bones."

"Her prayers were short and direct; simple yet comprehensive; the soul's sincere desires indited by the Holy Spirit."

"She always came from her closet to the hall for morning worship with her cherished Bible on her arm, or in her hand, with one finger between the leaves where she was about to open to us the Scriptures. We sang from Nettleton's Village Hymns. As the number was named we could see the glance of an eye that would know whether all had brought their books. If some had forgotten, she said nothing then, but perhaps in the afternoon she would ask us to read Deuteronomy and note the words 'remember,' 'observe,' 'take heed,' and see how God regards forgetfulness. Some of us then first realized that it may be a sin to forget."

"That old octavo Bible, with Scott's references, had been her companion from the beginning of her teaching in Buckland. At the time of her death it was in its third binding. She used to say she did not know what she should do when it needed a new dress, for the margin was already too narrow to admit of another trimming. She did not know that there was another with the same references in the United States; she had tried in vain to obtain one; one publisher had given encouragement of republishing it, but if he did not, she thought that when this was past use she should have to send to England for one of the same kind. Before that time came she had entered upon the glories it had revealed. We used to think if she should ever lay it aside for another, we should all esteem the old one a treasure. I am glad to know it has a place in Williston Hall."

It would almost seem that the writer of the following, though unaware of what others had written, had designed to recapitulate the preceding quotations.

"Some who have only heard of Miss Lyon as a woman of rare gifts and piety, have an idea that she was cold, distant, or sanctimonious. Nothing could be more untrue. In her natural temperament there was an overflow of animal spirits. But this was consecrated, and made to do good service in climbing many a Hill Difficulty in her life-journey. She was uniformly cheerful, often sprightly and vivacious. In the darkest day her motto was 'Hope thou in God, for I shall yet praise him.' Some think that she was lacking in refinement. So far from that being true,—as Dr. Brooks, her physician, has truly said,—'she was unusually refined; only her refinement was not acquired, but a part of her being, that grew with the growth of her other virtues.'

"We also think that Miss Lyon's personal appearance has been misrepresented. Her portraits all seem to us caricatures. She could not be shown on canvas. Her complexion was pure pink and white,—her eyes a clear blue, her hair a lovely shade of light brown, which waved over her forehead and temples. Beauty as well as strength dwelt in the sanctuary of her face. Sometimes we saw more of the strength, sometimes more of the beauty. Her countenance was a transparent mirror which reflected every shade of feeling. So perfect however was her self-poise, that she never exhibited petulance or anger. We saw her once under circumstances of great provocation. Tears stood in her eyes, but they never flashed with resentment. We have seen her look tired and grieved, never provoked. To watch her when worn with anxiety and labor, was to pronounce her face uninteresting, but it was quite another thing when matters of moment demanded prompt attention: as when consulting with Deacon Porter and the carpenters, surrounded by the timbers, rafters, and beams of the new building (in 1842). We noticed the same expression at the beginning of a 'new series' of studies, as she solved the complications that would now and then occur in the adjustment of the domestic and literary departments. These two must not collide. They never did. If a young lady recited in history at ten o'clock, she could have no work at that hour. Occasionally there seemed a tangle. With haste Miss Lyon would leave the platform and walk to the floor of the seminary hall. One arm might be slightly akimbo, one finger placed over her lips. Possibly her cap might be a little awry, although usually her dress was faultless. With dispatch she would call up the different classes. 'Miss Whitman, Miss Moore, please take these names.' Then as by magic, all the machinery would be adjusted smoothly for the next six weeks. Miss Lyon would quietly resume her seat, and perhaps proceed in her unique manner to correct some household evil. At one time she started off with a short talk on comparative anatomy. The naturalist, from one bone or one tooth, knows the entire animal and tells us whether it ate grass or flesh. So, single deeds indicate character. If a person does or fails to do some little thing, the whole nature is revealed. If Domitian would amuse himself by catching flies and piercing them with a bodkin, it was to be expected that he would kill Christians. One illustration followed another. The great principle was developed in a masterly way. It was so far an able disquisition, but we knew there was something practical to follow. It came at length. The descent was easy; it was also solemn. It seemed that much to Miss Lyon's satisfaction the ironing-room had been nicely refitted. But upon its first use the white sheets had been sadly discolored. Some showed the imprint of the iron, others had even been burned through. Miss Lyon did not care so much for the spoiling of the goods. She could take that joyfully. But it pained her that any of her dear family should evince a carelessness akin to recklessness. It was moral tarnish. It might be a straw but it showed the way of the wind. The lesson was effectual. Those words on character shown by little things, made many a heedless girl considerate and wise.

"In conversation with friends, Miss Lyon's face was very beautiful. So it was wont to be in her afternoon lectures. Her voice was sweet and strong. She was fluent but not flippant. Her sentences were well rounded. Many of her words had in them the weight of a talent. She was often jocose. She could use satire, though this was rare and only when the occasion required it.

"But how radiant was the face of our teacher when she opened to us the Scriptures! Then it shone like that of the man of God when he came down from the mount. She culled gems from the Mosaic ritual and pearls from the hard names in Chronicles; from David, and Isaiah, Hosea, and the Apostles she brought forth treasures new and old; so, 'beginning at Moses and all the prophets, she expounded to us in all the Scriptures the things concerning' Jesus Christ. And our hearts burned within us.

"We have often attempted to analyze the secret of her influence over us. She had great mental resources. No one could resist the impression that these were deeper and richer than we had fathomed. But the sense of her reserve power was not the key. One element was her deep interest in her pupils. She was not demonstrative in her affection. But as her beaming face looked down upon us, as those speaking eyes met ours, every one, even the most refractory, felt that she sought our highest good, and more, that she loved us after the manner of our mothers. Miss Lyon's was the warm, glowing, motherly heart for every scholar in the school. No matter how many they were, the names of all, and especially of those out of Christ, were graven on her hands. Another element of this ability to influence, was the honesty and intensity of her convictions. Still another, was her utter unconsciousness of self. Her great thoughts stood foremost, while she was hidden behind them. Her scholars sat in those seats before her and were not simply swayed as the trees of the wood; they were molded over, changed radically and permanently in habits and character. She was a mighty moral architect. As she spoke, the old would crumble, the rubbish be removed, solid foundations be laid, and a comely edifice begin to rise. The Holy Spirit wrought this spiritual rebuilding. She was in communion with him, ever relying on his aid. In prayer was the hiding of her power. She asked what she would, and it was done unto her."

"The Lord raised up Miss Lyon for an important work. That work she nobly accomplished, and entered into rest. Then to him let us give all the glory; and yet even Christ himself said that the devotion of the woman who poured the ointment on his head should be told for a memorial of her wherever his gospel should be preached."

CHAPTER X.

GROWTH UNDER MISS LYON.

1837—1849.

SOME of the friends of the seminary who had admired its principles had doubted their expediency, and also feared that pupils could not be procured without greater latitude in the terms of admission. But the result showed the wisdom of the effort to secure the best school, rather than the greatest numbers. Some of the first year's pupils had been admitted for only a part of the year. Others were waiting to take their places, and the first catalogue records the names of one hundred and sixteen. The second year one hundred were crowded into the building, four hundred having been refused for want of room. The same reason kept away hundreds the third year, and though accommodations have been increased from time to time, and the standard of admission has steadily been raised, Mount Holyoke has never been without hundreds of applicants, even when in later years, other institutions have been springing up almost at her doors. Yet there have always been obstacles to meet.

It illustrates the prevalent opinions of the time, that when Dr. Anderson was invited to make the anniversary address in 1839, many friends, solicitous for his reputation, sought to dissuade him from indorsing such an enterprise. In an association of gentlemen, most of whom became distinguished, only Prof. Bela B. Edwards favored his accepting the invitation. Some of the ideas in that address reflect the views of the wisest men in that transition age in the history of the higher education of woman. He said:—

All of the education due to the sex has not been given. We are somewhat influenced by ancient prejudices. We have misgivings as to the effect of a liberal education upon women; as if learning disqualified rather than fitted them for their appropriate sphere. Yet there is progress. Education for woman is not now what it was in our own youthful days. It is in a transition state. Whether experience will lead to a three or four years' course of study is not yet known. This much, however, is certain, there are important experiments in progress, and this seminary is one of the most important. Should it prove that two hundred young women can be retained in a seminary during a three years' course without on the one hand restricting the well disposed too much by rules made for the wayward, and on the other without those corrupting influences to which large seminaries are supposed to be liable, the founding of this institution will form an epoch in the history of woman in this land. As a necessary experiment, made in the best section of our country, and under the management of sound common sense and piety, it has had from the first our cordial approbation. And its development thus far has increased our hopes. It is not a local institution. The experiment is one of general interest, and its success will be a national good. It is not enough for objectors to argue that there are defects in the plan or in the principles which lie at its basis. They must think out something better. They must show that if this institution did not exist there would be something equally good for those in whose especial behalf it was designed. I have not yet seen reason to believe that this can be done. Coming into existence in that state of the progress of female education when everything is in a changing state, the institution can hardly be perfect. Should experience suggest improvements, they will, no doubt, be made. One thing is thankworthy; it takes the girl at the point where her education was regarded finished in the last generation, and requiring more preparation for admittance than made up that education, it employs her mind for three continuous years in that way which experience shows is most salutary in its influence upon mind and heart, upon character and life. Should it fail here, it will probably be because the public mind is not prepared for so great an advance, or else because of some essential departure from its present course. But if it should fail, we may be sure that from its ashes another would rise phœnix-like, more perfect because of the experience this had acquired.

It is cheering to cast our thoughts forward and see what would follow should such education become general among women. It would add to the sum total of efficient mind. It would exert an influence upon every profession in life and on every department of society, upon our religious character and all our social and civil institutions. It would be a powerful transforming influence, touching all the springs of action and reaching all the fountains of enjoyment.

Let the grandeur of the object encourage the founders of this seminary. Let what has been begun in prayer and faith and been crowned with such signal spiritual blessings, be finished as soon as possible. Yonder convenient edifice, rising on one of the healthiest locations in the land, and looking forth on some of the choicest scenery in nature, needs to be completed. Its grounds, well chosen for retirement, ought to be laid out so as to invite the pupils to morning walks in the fresh air. And why should not our daughters, who can comprehend the wonders of natural science quite as well as their brothers, have the advantage of a complete philosophical apparatus and a well assorted library? I know that these things enter into your plans and that only the want of funds occasions delay; but I feel assured that the needed funds will be forthcoming as soon as the enterprise is understood. Regarded merely as an experiment it is worth all it will cost. But as an institution destined to bless generations to come and be the commencement of a higher order of seminaries for young women, it will have peculiar claims on the beneficence of the intelligent and the good.

Rev. Dr. Mark Hopkins, the next anniversary speaker, said, referring to the seminary: "Many judicious persons still look with suspense at the issue of the experiment; though I am happy to say that so far as I know, the objections are vanishing as the institution progresses and becomes better known; and what I have seen to-day strengthens my conviction that those objections will vanish entirely."

Prof. B. B. Edwards said in the fourth annual address: "If a course of study like that pursued in this seminary could be introduced into every state it would be one of the firmest props of the Union. No disorganizing influences emanate from it. No beetle-eyed prejudice, no narrow-minded bigotry, can find a home where the sciences are truly taught; the air which is breathed is too invigorating; the impulses which it prompts are too noble. We do not deny that there are possible evils connected with a protracted and public course of education for woman. We think, however, that they can be obviated by a due measure of forethought and care on the part of guardians and teachers."

At the twenty-fifth anniversary, Dr. Anderson took public occasion to say that he never had seen reason to regret delivering his address, though contrary to the advice of friends; that he became more and more impressed with the highly practical character of Miss Lyon's plans; that though they then seemed large, they appeared more comprehensive afterward, and time had indorsed them all; and that he hoped the trustees would be slow to allow any change. These words show that the speaker twenty-five years ago was satisfied with the experiment made by the pioneer institution. Yet many are still in doubt. The seminary was more than forty years old when T. W. Higginson wrote in "Common Sense About Women": "Why is it, that whenever anything is done for women in the way of education it is called 'an experiment,' while if the same thing is done for men its desirableness is assumed, and the thing is done? Thus, when Harvard College was founded, it was not regarded as an experiment but as an institution. The 'General Court,' in 1636, 'agreed to give four hundred pounds towards a *schoale or colledge,*' and the affair was settled. Every subsequent step in expanding educational opportunities for young men has gone in the same way. But when there seems a chance of extending some of the same collegiate advantages to women, I observe that some of our periodicals, in all good faith, speak of the measure as ' an experiment.' "

These facts make it seem less strange that there should be difficulty in raising funds for such an enterprise. The wonder is that Miss Lyon succeeded so well when the general public did not yet take interest in the experiment. She had been compelled to open the school with room for only half the pupils in her plan. "We will begin," she said, "and when it shall be seen what is the manner of our building, it may be that the wise-hearted will give us of their treasures till we can finish the house of the Lord." She had begun. She was no longer obliged to speak merely of what she intended, but could state what the school was doing and add facts to her plea for means wherewith to complete her plan.

Before the close of the second year she prepared a circular which set forth the ideas embodied in the seminary, invited friends to examine their operation, and confidently asked for twenty thousand dollars to finish the buildings and five thousand more for furniture, library, and apparatus, closing with these words: "Are there not many who will remember this cause around the mercy-seat of him whose is the silver and the gold? 'Except the Lord build the house, they labor in vain that build it; except the Lord keep the city, the watchman waketh but in vain.'"

The seminary had secured some hold on the New England heart and many 'wise-hearted' gave willingly. Yet Miss Lyon had use again for all her powers of persuasion. Even among her friends there was opposition to enlargement. At the quarter-centennial reunion Rev. Dr. Joel Hawes of Hartford said: "I have rarely known a person so successful as Miss Lyon in winning others to her views. This was because she had good common sense and a good heart. I well remember my early objections to the system which she proposed. I remember too how I presented them to her at the house of Dr. Hitchcock, and how she met every one, and I said that day, 'Miss Lyon has converted me.' I had objections also to enlarging the building, but was again converted to her views. As we look around, we feel that God has not only fulfilled his promise but has let none of the words of his handmaid fall to the ground."

At her urgent solicitation the trustees had already begun to plan. They voted in April, 1839, "to contract for brick to add thirty feet to the building," and three months after, "to procure plans and proceed to erect an addition of at least fifty feet, as soon as funds justify." A year later they voted, "to extend the present building about seventy feet to the south, and to erect a wing on the end when extended, that with a corresponding wing on the north end shall be sufficient in all to accommodate two hundred pupils, according to the original plan," again repeating the condition, "provided in the judgment of the building committee, the funds will justify." The treasurer was allowed to hire a sum not exceeding five thousand dollars. The first building committee had been continued in office against this time of enlargement. The south end had been built without windows, save one at the end of the long hall in each story. For three years this unfinished appearance had borne silent witness to Miss Lyon's faith in the future addition. Work upon it was begun in the autumn of 1840. It was seventy feet long and of the same width and height as the original structure. From its end a wing forty feet wide extended eastward seventy-five feet; but there were no funds for a north wing.

As at the first, there were trying delays, and the new part was not finished till December, 1841, two months after the term began. Again ladies' benevolent societies had contributed of their handiwork, and pupils who could not themselves return, aided in obtaining what was needed for their successors. Miss Lyon told the crowded one hundred and seventy that they could be very happy in close quarters for a time, with a prospect of such ample room the rest of the year. One says: "Already the seminary had a history and we were often entertained with reminiscences of the past, and Miss Lyon herself gave details of the enterprise from its first inception as an idea in her own mind. She used to tell us of her gratitude to the pupils of the first year who stood shoulder to shoulder with her in bearing the burden, when everything was unsettled, and that we were privileged with a discipline like theirs—a blessing which those of the intervening years could not know." "We were so full of enthusiasm at entering Mount Holyoke Seminary," says another, "that we did not mind inconveniences. That we did not have all we needed was not Miss Lyon's fault. Her heart yearned to do more. It is wonderful that with so small means she could do so much."

As before and always, Deacons Safford and Porter stayed up her hands. Another writes: "My earliest memory of those noble trustees was of seeing them mounting the long flights of stairs, in some emergency, bearing on their shoulders the young ladies' trunks. I think I had something of the feeling of the apostle when he said to his Lord, 'Thou shalt never wash my feet.' It was not many years before baggage elevators performed that service; and now I hear that a passenger elevator makes fourth-story rooms the most popular in the house. I might depict primitive scenes, like trying on a winter's morning at five o'clock to ignite, over the whale-oil lamp, a coal with which to kindle a fire in the little open stove. And yet I think those early pupils were as happy as any of later days."

When the new part was ready for use, a note book records: "Miss Lyon took her subject one morning at prayers from Ezekiel xliii. 12. 'This is the law of the house; upon the top of the mountain the whole limit thereof round about shall be most holy.' The thought was one she often dwelt upon,—that the place was holy because it had been built with money dedicated to the Lord, and that life in such a place was full of responsibilities and new obligations."

In December Miss Lyon wrote: "We have a valuable addition to our building. I suppose more than fifty thousand dollars has been expended in all; the trustees are somewhat in debt, and something more must be laid out in finishing. One more addition will complete the design. When that will be done I know not."

She did not live to see it. Even the finishing of the part now erected was not completed during her life. Besides the piazza of two stories, it was designed to relieve the plainness of the structure by an observatory on the roof. The engraving published in Miss Lyon's memoir represents this feature of the plan, though then existing only on paper. When the north wing was built in 1853, the committee in charge was authorized also "to erect an observatory on the center of the main building, as originally intended." But the funds were insufficient and it was not done till 1860, more than ten years after Miss Lyon's death.

Delays never discouraged her nor slackened her zeal; and however it might be with outward enlargement there must be steady growth within. In no other way could the seminary answer its end. "Progress," no less than "Holiness to the Lord," was written on every brick. In 1835 she had said to the public that the literary standard of the seminary would be as high as that at Ipswich, and like that it would also be progressive. True to this idea the requirements announced in 1837 were in advance of those at Ipswich in 1835, although embracing essentially the same studies. At Ipswich it was not expected that all would take the regular course. Any not under fourteen could enter for primary studies, and when prepared, be admitted by examination to either of the two regular classes. At Mount Holyoke there were to be three classes, none were to be received under sixteen, nor without examination, and the studies required were to be taken in order. Requirements for admission were: "An acquaintance with the general principles of English Grammar, a good knowledge of Modern Geography, History of the United States, Watts on the Mind, Colburn's First Lessons, and the whole of Adams' Arithmetic." Other studies however advanced could not be substituted for these. The course was arranged as follows:—

JUNIOR CLASS.

English Grammar, ...	Murray
Ancient Geography, . . . Worcester's Ancient Atlas
Ancient and Modern History,	Worcester's Elements, with Grimshaw's France, and Goldsmith's England, Greece, and Rome.
Political Class Book, ...	Sullivan
Botany, ...	Beck
Rhetoric, ...	Newman
Geometry, ...	Playfair's Euclid
Physiology, ...	Hayward

10

MIDDLE CLASS.

English Grammar, continued,	Murray
Algebra,	Day
Botany continued,	Beck
Natural Philosophy,	Olmsted
Philosophy of Natural History,	Smellie
Intellectual Philosophy,	Abercrombie

SENIOR CLASS.

Chemistry,	Beck
Astronomy,	Wilkins
Geology,	Mather
Ecclesiastical History,	Marsh
Evidences of Christianity,	Alexander
Logic,	Whately
Moral Philosophy,	Wayland
Natural Theology,	Paley

Butler's Analogy.

It was also stated: "The studies of each class are designed for one year, though pupils will be advanced from class to class according to progress and not according to the time spent. Some may devote half their time to studies not in the course, Latin for instance, and take two years for the studies of one class. Reading, composition, calisthenics, vocal music, and the Bible will receive attention through the course. Those who are deficient in spelling and writing will study these branches whatever may be their other attainments. Every one should be supplied with a Bible, a dictionary, and an atlas. Those who have a concordance and commentaries on the gospels are requested to bring them. The Bible lessons will begin with the New Testament."

Miss Lyon attached much importance to vocal music. As early as 1832 she wrote: "When passing near the music room last summer and thinking that probably a large part of the choir had no more natural ability for singing than myself, I found it needed grace to restrain a rising murmur. I have sometimes felt that I would have given six months of my time when I was under twenty, and defrayed my expenses, difficult as it was to find time or money, could I have enjoyed their priv-

THE MINERALOGICAL CABINET.
Lyman Williston Hall.

ileges." Deploring her own lack, she gave all her Holyoke pupils who could sing, the opportunity for instruction in graded classes.

The outline of study was so well matured that Wayland's Political Economy and Milton's Paradise Lost, introduced in the second year of the seminary, were the only English studies added until 1855; but the annual catalogues increased the requirements in mathematics, and show that the standard in other departments was steadily rising.

Miss Lyon never lost her early enthusiasm for the natural sciences. Believing the God of nature, of providence, and of grace to be the same, she traced his hand alike in history, in science, and on the inspired page. She was never afraid of scientific revelations. "If the Bible," she said, "only take the lead in our schools, I care not how closely the sciences follow." But the catalogues give only glimpses of progress here, no pains were taken to set it forth. She who, when learning to write, returned her copy set in Latin asking for it in English, lest some reader might think her wiser than she was, not only avoided every statement which might make the seminary appear better than it was, but gave only modest representations of what was really done. No new advantage it offered was ever announced until it had been put in successful operation. The seminary journal tells of lecture courses given in the third year, on architecture, by Prof. Snell; in 1844, on galvanism, by Dr. Hitchcock; and also by Dr. Hitchcock on anatomy and physiology, illustrated by a manikin. To make these lectures most profitable the school was divided into classes for recitations upon them. July 2, 1847, Prof. Snell helped unpack and arrange new philosophical apparatus, and began his course upon natural philosophy, having been preceded by Rev. Mr. Stone on elocution. As one's eye runs over the journal it is caught by a brief notice of "a lecture on temperance given here May 7, 1846, by a Washingtonian of some note in the last three years."

The writer adds, "He is not yet thirty, but if he lives he will soon be known throughout the length and breadth of the land. His name is John B. Gough!"

Outside the course, provision was made for all who desired Latin, French, drawing, or piano practice; the only extra charge being for use of piano.

French was taught from the opening of the seminary. For nine years the teacher was Miss Moore, afterward Mrs. Burgess, a senior and an assistant pupil of the first year. A part of the time her work was supplemented by a native of France who came regularly from Northampton to give lessons. Most of the time since then the teaching has been by a French or German lady resident in the family.

There were classes in Latin every year after the first. In 1840 Miss Mary M. Stevens came, as an assistant pupil, to teach this branch, of which she continued in charge for eight years. The third annual catalogue speaks of a four years' course in contemplation, that Latin might be included. A year later the remark is added, "But it is supposed that the views of the community will not at present allow of it." Though optional, more than one-fourth of the school were already pursuing it. For the next five years this study was "earnestly recommended by the trustees and teachers, not only for the knowledge itself, but also as a means of gaining discipline for the higher English branches." The catalogue of 1846 states: "A large proportion of the late senior classes have had some knowledge of Latin, and it is believed that the state of education in the community is now such that it can be required hereafter of every graduate." It was accordingly placed in the course. The amount was small at first, but increased from year to year. The next year it was made a preparatory study. The class that entered in 1848— the last candidates Miss Lyon received — were required to have a good knowledge of Andrews and Stoddard's Latin Grammar and Andrews' Latin Reader. Miss Lyon continued to press the matter of a four years' course, but we learn from a private letter that the trustees were still "afraid to venture it. She wanted time for Greek and Hebrew and more music. She often said she thought the time would come when Bible class teachers would feel that they must study the Scriptures in their original languages."

Some have said that Miss Lyon builded better than she knew, and that if she were to visit the seminary at the end of its first half-century, she would be surprised at the advance on every side. But those who best understand the scope of her plans reply that though she may not have seen all that it would require, yet her ideal, as described in the circular of 1835, was that this "school for Christ" should "be furnished with every advantage which the state of education in this country will allow."

"With almost prophetic eyes," writes Miss Spofford, "she portrayed the future of Mount Holyoke Seminary. She dwelt upon what Yale and Harvard were in their beginning, and declared her conviction that the seminary of her day was to that of the future only as the germ to the full grown plant—to the tree under whose increasing shadow should be taught and given to the world, minds whose breadth, resources, and attainments would sink the meager acquirements of our day into insignificance. She did not call the school a college, for she had a wholesome aversion to anything like pretension, but she proceeded with all the energy and wisdom of the great woman that she was, to make it as much of a college as was possible in her day."

CHAPTER XI.

GROWTH UNDER MISS LYON—CONTINUED.   RELIGIOUS HISTORY.

1837—1849.

IT is so often true, as Henry Martyn said, that "Christ is crucified between two thieves—classics and mathematics"—that many doubt whether a high standard in piety can be maintained along with a high standard in scholarship. It is said that Jonathan Edwards could make the solution of a question in metaphysics almost as powerful a means of grace as prayer. On the other hand, Miss Lyon found that Christian principle secured a higher standard of study and recitation than any other motive could inspire, and believed with the wise man that "the fear of the Lord is the beginning of wisdom." The history of the seminary has always accorded with the maxim, "to have prayed well is to have studied well."

Year by year there was steady spiritual growth; for there was devout planting and watering, and God gave the increase. In the words of Dr. Hitchcock: "It was an almost uninterrupted display of divine converting power. And yet so busy and enthusiastic in literary instruction were Miss Lyon and her teachers, that one would hardly have thought of the existence of that deep under-current which seemed to flow from the river of God and refresh the whole landscape. But the current was always there and thence came the power that kept the windows of heaven always open." Hundreds began the Christian life and Christians grew in grace. Sometimes a more special work was sudden and rapid. Sometimes it began the first week of school and continued through the year. If the blessing tarried, perhaps Miss Lyon would withdraw for a few days to her quiet room at Deacon Safford's or Deacon Porter's, to wait on the Lord and learn more perfectly his will. Her own unction would be communicated to the teachers on her return and soon the presence of the Spirit would be known in the household by the subdued voice, light footstep, and gentle manner of all; by the chastened spirit of Christians; by deep conviction of sin and great joy in Christ the Saviour. The human instrumentalities were earnest prayer and the plain presentation of divine truth. Except on fast-days, ordinary pursuits were neither suspended nor interrupted. Hence there was little or no reaction.

The following from the seminary journal shows how the annual fast-days were observed. This was on the first Monday of the year, but with change of subjects for prayer, the description would apply equally well to a February fast-day. "The dawn found a few unitedly asking for a blessing on the day. Instead of going to breakfast many remained alone in their rooms. A prayer meeting was held at eight o'clock, and at nine we met in the seminary hall. Miss Lyon read John vii. 37. She seemed to feel that this may be to us 'the last great day of the feast,' as our term is soon to close; and her earnest, impressive appeal to all 'to come and take of the water of life to-day,' will not soon be forgotten. She then gave us subjects for thought and prayer. She asked us to pray for the conversion of all in this family, that so Christ's kingdom may be built up and the world converted; for all missionary societies, especially for the American Board, and for its officers, on whom such a crushing weight of care and responsibility rests; for the different missionary stations; and for all missionaries, particularly those that have gone out from us; to pray for them by name; to pray that more laborers may be raised up, and that the missionary spirit may be increased all over our land. She talked to us nearly an hour, and made us feel that the day would be too short for the petitions we should desire to present before the mercy-seat. Prayer meetings were appointed at different hours through the day. Roommates, by attending these in turn, had their own rooms, alternately alone. The prayers were all voluntary and those who offered them appeared to draw very near the throne of grace. The Holy Spirit seemed specially present. At four o'clock three meetings were appointed; one for professing Christians in the senior class, another for the same in the junior and middle classes, and a third in Miss Lyon's room, for such of those without hope as were disposed to attend. Two only of this number were absent. At the close of this meeting Miss Lyon invited any who desired to be made special subjects of prayer to come again at seven o'clock, and requested several of the teachers to meet with them at that time. More than twenty returned. At half past seven, professing Christians met in praying circles as is usual Sabbath evening."

In the first year of the seminary there were but ten or twelve who did not class themselves as Christians when they entered. The first marked revival occurred in the second year. The last Thursday in February had been spent in fasting and prayer for literary institutions. Saturday was a day of recreation, but nearly the whole school gathered in a prayer meeting in the afternoon. When it was closed no one rose to go. After another prayer it was proposed that each should retire to her room for half an hour, and that then those who wished should come again. They went and returned to plead once more for their companions out of Christ. The next day fifteen were rejoicing in hope. Gradually their number increased to thirty, and at the end of the year only one remained without hope. Not long after, prayer for her was answered also. The effect of that revival was felt for years.

A pupil writes of the next year: "It was our third Saturday evening in the seminary. It had been recreation day and we had enjoyed what Miss Lyon always said she wished us to enjoy, lively, social intercourse. But at our evening meal there was one table where there was unbecoming mirth and even frivolity, and most of those at the table were professing Christians. As we sat back for family devotions, Miss Lyon rose and said: 'I have thought that perhaps some of God's dear children here do not know that his Spirit is striving with souls in our family. There are those who long to find a Saviour before this Sabbath has passed. Would you not like to pray for them?' She said no more. Half-formed purposes to seek the Lord were strengthened and careless Christians felt that they were on holy ground. As we passed out, a stranger threw her arm around me saying, 'You do not know me, but I thought perhaps you would pray that I may find Christ this Sabbath day.' I had been reproved by Miss Lyon's words, and now another reproof had come. I sought my closet to weep and to feel that if a child of God, I must always be ready to pray and to labor for souls. Many others left that room with similar feelings, and those few words of Miss Lyon seemed given to guide us all the year, and we know that they have led many to watch carefully all their lives lest by any means they grieve the Spirit of God from an inquiring soul."

It was that year that plans were formed for enlarging the building. Exceedingly desirous to have this done, and striving also for improvement in the department of instruction, Miss Lyon feared lest becoming engrossed in these things she should lose a spiritual blessing. God knew her motives and granted her desires. While other interests prospered she had the greater joy of seeing all her children walking in the truth, thirty of them having entered that path during the year.

The future was bright with hope. But at this point, to use Miss Lyon's words, "God saw that great trials must be set over against great prosperity." In the summer vacation of 1840, one after another of her pupils was attacked with fever, and nine of them died.

On the last Sabbath evening of the term, in the usual circles for prayer, there had been given a pledge to remember each other at sunset on the Sabbath, when they should be separated. "Three weeks from that hour, one on whom heaven was opening prayed for the others and then said: 'Mother, they will pray for me now. They will not know that I am dying, but I am so happy to think they will pray for me.' Another, as she asked the hour, said, 'I should be glad to go while they are praying for me;' while a third said to weeping friends, 'There will come peace, for they are praying God to bless me.' One was released at that hour of prayer and others gathered strength soon to follow, while the Lord said to some longing to depart, 'Ye cannot come to me now, but grace shall be given you to meet the ills of life as you go back from this view of heaven.' Fullness of joy in Christ was bestowed on all who died, and to those who came back to earth an experience of love and trust in him such as they had not known before." So wrote Fidelia Fiske, who was one of the number. But while departing ones rejoiced, all the waves and billows seemed to be rolling over Miss Lyon. "I was afraid," she said, "to take up a paper lest I should see some new name added to the dead and not have strength to meet it. I was afraid to ask a question, or even listen to conversation, lest I should not find myself prepared for what I might hear. There were days in which I could not attempt anything except to ask God to hold me by his own hand. I had no heart to ask for anything but to have my trust in God made strong." To another she said: "The hand of God has been laid heavily upon me. I have been led through deep waters, but they have not overflowed me. None but my heavenly Father knows how great a trial this has been. While others have been inquiring about the natural cause, I have felt that we ought also to inquire about the moral cause, and seek to know what the Lord would have us learn." She was always as willing to learn as a child and ready to make any change in her favorite plans that might be required. In allusion to false reports she said: "If we are grieved by what is unjustly reported, let us remember the example of our Saviour. He opened not his mouth. Let God in his own way and time vindicate his own acts. Let us commit ourselves to the covenant-keeping God, who doeth all things well."

The fever caused many failures among the candidates for the next year, but others were waiting for vacancies and the house was filled. She received them with peculiar tenderness. One who met her that day for the first time, tells us of the opening, October 3rd. "We met in the hall at nine o'clock. Miss Lyon read a psalm and led in a prayer which was very comprehensive and affecting. Its burden was that if any of us had not given ourselves to Christ we might be led to do so even before we entered upon our studies; that the blessings of life and health might crown the year, if consistent with God's will, but if any were to be removed by death, that they might here ripen for heaven."

In the midst of her sore trial her thoughts turned more to the goodness of God than to any other theme, and this was the subject for the Bible lesson of the first Sabbath. Alluding to it at the breakfast table she said: "We hope to spend forty Sabbaths together, and will not those who love the Lord speak often one to another? then will he write for us a book of precious remembrance, for he has said he will. After forty Sabbaths we shall separate; we have dear ones here who have no hope of meeting us beyond this life. May not such a hope be given to some of them this first Sabbath morning?" Another morning she used the text, "In everything by prayer and supplication with thanksgiving let your requests be made known unto God," and added, "when he says everything, he means just what he says. Those who are Christians can do so, and those who are not, may begin to-day. Do not be afraid because you know you are not a Christian. Does the heart of one go out tenderly this morning to the friends she has left? Does any one feel anxious about her examinations? Carry all to your heavenly Father. You may thus get an acquaintance with him that will make you long to say to God, '*my* Father.'" Many of her pupils recall words like these among her first addresses: "My heart goes out very tenderly this morning to those parents who have entrusted you to my care. They have no choicer treasures than these precious daughters. We are ready to labor for you in love and fidelity, and may you all be faithful. And oh! what inexpressible tenderness in the thought that you may all be preparing for heaven here!"

The blessings of life and health, bodily and spiritual, were bestowed. That year there were thirty who at first stood aloof from Christ. One of them came to him during the first week. At the opening meeting for Christians, Miss Lyon reminded them that they were enjoying privileges which they could not buy—that a debt was due to the founders of the seminary which could be canceled only by a useful Christian life. Referring to those who had not found Christ she asked, "Shall we help them find him?" Her appeal met with a warm response, and Christians labored diligently for souls and helped care for the lambs as they were brought into the fold. To these co-workers we find her saying, May 19th: "It is a solemn thought that all that we do for the four in our family still strangers to Christ will be done in a few short weeks."

Perhaps no morning exercise of this fourth year was more impressive than that of December 3rd. Miss Lyon was passing through new trials. A few weeks before, she had been called to the death-bed of her youngest sister, for whom she had a strong affection. Now she had come from watching by her dying mother. When she entered the hall, looked around on her pupils and gave them her good morning, there was an irresistible drawing of hearts toward her. She was usually so busy and said so little of personal matters that few realized the strength of her love to her friends. Controlling her emotions, she made no effort to conceal them, as she spoke tenderly of her mother, to whose comfort she had hoped to minister for years to come, whose counsels she had always loved to seek, whose daily prayers she had never known what it was to miss before. And now she felt a loneliness of which she said she had no previous conception. Then she turned to the words of Paul, who would not have us ignorant concerning them who are asleep, nor sorrow as those without hope. With her eye upon new visions of the things that God has prepared for them that love him, she had returned to her beloved flock. They knew before she told them, "Heaven seems nearer to me than ever before, and labor on earth sweeter"; and they saw new force in the words, "Our first great blessing is that we may be in Christ, the second, that we may labor for him."

These repeated strokes bore heavily on Miss Lyon's strength. She met the school as usual for religious instruction, but others took her place in the class room much of the year. New vigor was derived in the early summer from a journey with Deacon and Mrs. Safford, from which she returned to find encouraging progress on the new part of the house. There was almost perfect health in school, and there had been so rich a spiritual blessing through the year that she said, "It seems to me that I never had a school in which there was more of the spirit of heaven. All but three express hope in Christ."

Many who had been detained by the illness of the preceding autumn were present at the re-opening in 1841. Taken from their books, they had been learning other lessons in their prolonged vacation. The Sabbath twilight concert had bound them to each other and made more dear the privilege of united prayer. It was on their return that the daily "recess meeting" began, and the fifth year was one of marked growth in Christian character. "About forty who came among us strangers to God," wrote Miss Fiske, "were numbered, we hope, among the followers of the Lamb before they left us."

The following description, by Miss Whitman, is as applicable after forty-five years as it was then. "The recess meetings are short seasons of prayer held at eight o'clock. Often at the ringing of the bell the whole school is seen moving toward the appointed places. For about two minutes the passage halls are thronged. Then all is silent save the voice of song from eight or ten different quarters. The low voice of supplication follows. The bell again rings. All rise and return to their rooms. A few words are exchanged between roommates, and all are again absorbed in study. Refreshed in spirit and strengthened in mind by this elevation of the soul to heaven, they are able to re-apply themselves vigorously to their lessons. The moral influence of these little meetings has always been great. Those who are disposed to continue in their rooms, cannot be rude with such impressive voices around them. Every influence leads them to go with those whose hearts incline them to pray. The section teacher seeks to induce all the members of her own division to attend. As the meeting is entirely voluntary, the young ladies regard it as peculiarly their own. It is a thermometer, indicating the spiritual temperature, both among Christians and the impenitent."

In the work of organization, Miss Lyon had been relieved of much care in the literary department by her associate principal, Miss Caldwell, who left the seminary at the end of the first year; she soon became the wife of Rev. J. P. Cowles, and with him, for more than thirty years, carried on the seminary at Ipswich. For the next four years Miss Lyon had gone on without a nominal associate. Her band of teachers had been reenforced the second year by the three graduates of the first class, and the third year by four of the twelve who formed the second class. Some of these having gone to other fields, six new ones came to Miss Lyon's aid in

1842. Miss Abigail Moore, a niece of Miss Lyon and a graduate the first year, with Miss Mary C. Whitman, of the second year's class, were made associate principals. So fully were they able to carry out her plans that they seemed scarcely less essential to the growing seminary than Miss Lyon herself.

In December she wrote: "We have had a very prosperous year in worldly things. Everything is systematized, and Miss Moore and Miss Whitman urge forward the wheels so successfully that all seems more than ever like clock-work. I enjoy very much having everything done better by others than it can be by myself. If this pleasure continues to increase as it has done for a year or two, I hope I may be prepared to be happy in being old and laid aside as a useless thing. But in spiritual things we are less favored. There has been less interest than we have had any year since the first. Pray for us that we may not receive all our good things in this life." In a few months this prayer was answered in a way they had not known before.

The five previous years had borne rich fruit. The Lord had set his seal on the seminary. Out of one hundred and forty who had entered it disclaiming personal interest in Christ, over one hundred and twenty-five had enlisted in his service, and each year sent forth an increasing number of Christian laborers. It will be remembered that a prominent object of the seminary as first announced was "to prepare teachers for the millions of our population calling for education," especially "in the great valley of the West."

Teaching was the only calling, outside the home, then open to women, and there was scarcely a thought of going from the home land to teach. Four had become wives of foreign missionaries, one of them having left the first year to go to the Zulus in South Africa; a fifth was teaching in the Cherokee Mission at Park Hill, Arkansas. Eleven of the twelve who graduated in July of the second year were teaching in five different states before November. Nearly sixty had completed the course and many of them were filling important positions. The seminary was practically doing the work of a normal school. If in any respects it was less intense for a technical purpose than that which began at Lexington in 1839, it was deeper and more comprehensive in its design, since it was at the same time fitting women for all the relations of life.

Miss Lyon was grateful for the prosperity of the seminary and for all that had been done through it, yet she feared it was not doing all that it could for the kingdom of Christ. Her chief desire was not that the seminary should flourish, but that his cause should be promoted through it. She could accept no standard but the highest. Dwelling on the sacrifice of Christ, offered not for one person nor for one people but for all men, she said: "If we would labor aright for Christ our hearts must take in the whole world." She saw that this habit, like any other, must be gained by present practice, not by resolutions for the future. But how should school-girls be trained to live for the conversion of the world? "There is not a day," she said, "in which I do not ask how I can enlighten the understanding and direct the feelings of my pupils aright on this great subject. The sacrifice in giving my little is nothing in comparison with my anxiety on this point." After the second year a meeting had been held once or twice every month with this object—"to disseminate information throughout the school relative to the moral and religious condition of the human race, to excite inquiry, and to awaken zeal in the work of the world's conversion."

A quickened Christian life had already enlarged their desires. In their recess meetings, Monday was devoted to foreign missions, Tuesday to home missions, Wednesday to the Bible and Tract Societies, Thursday to home churches and friends, Friday to their own and other seminaries, and Saturday they pleaded for Abraham's seed, and prayed, "Thy kingdom come in all the world." Their petitions were specific as well as broad, mentioning by name the laborers in the different fields.

With Miss Lyon, praying and doing were never divorced. "We will pray," she said, "but let us also do, and do now. By waiting you may lose the little desire you have. Feeling without action is exceedingly dangerous. Give your cordial support to those societies whose great object is to save the world." Giving was made a personal matter. "I would not rouse your feelings merely," she said, "I would awaken your consciences. There is a standard of giving for every individual. And this we are to find out, each for herself. If it were written on the walls of our rooms how large or how small a sum we should give, we should not be treated as moral agents. God has a plan for every farthing he has placed in our hands. If we are willing and obedient we may know his plan; but no one will know how much he ought to give unless he has a strong desire to know. God will make our treasures, whether few or many, a touch-stone; a test of the willingness of our hearts. The Bible is our statute book, and when it makes known our duty we are not to answer again. If God asks a part of our pittance, we must not inquire how we can get along without it. We must not be careless of what we have, but remember that God's blessing depends on the manner we use what he has committed to us for his cause. The Bible teaches us to give a portion of our income to the Lord, and we must give it before we expend anything for ourselves. It seems probable that the Jews gave at least four-tenths of their income. Shall we under the gospel dispensation, with increased light and ability, do less? Our standard must be different from that of those who have gone before us. We ought to rise as much higher than our parents as we are younger, for we have more light and greater opportunities. Let us never go back one step but rather onward to the day of our death. If our parents or friends are not benevolent, let us seek to supply their lack of service. This contributing is the current money of the heart. It shows how much we love; and what a privilege, by giving money, to show our love for him who has redeemed us!"

A series of morning talks upon this subject closed with these words: "Before we take up our contribution, let us all take time in our closets to consider the worth of a single soul. When we tremble in view of the possibility of being lost ourselves, then let us do what is assigned to us for other souls. Have we ever given and toiled and prayed for those in darkness till we felt the sacrifice? Are you ready to go yourself to the ends of the earth for the salvation of others? If we send others to endure the toils, shall we not practice self-denial? If you desire in agony of heart to know what you ought to do, then give as the Lord shall show you. I often look forward to the day when we shall hear it said, 'Inasmuch as ye did it not unto one of the least of these.' Let us do it unto Christ, first of all giving to him our own hearts. I seem sometimes to look out through the crevices of my prison-house and see something of the work given us to do here. And you may see more than I see and do more than I have done. I love to think that my precious daughters will do for Christ's kingdom what I have not done and fear I never shall do."

At the meeting of the American Board at Norwich, Connecticut, in the autumn of 1842, Miss Lyon was deeply impressed with the thought that the seminary should be more thoroughly pervaded with the missionary spirit. Calling a meeting of her former pupils who were present she told them that the seminary was founded to advance the missionary cause, and that she sometimes felt that its walls were built from the funds of the missionary boards, for much of that money would otherwise have been cast into their treasuries.

At that meeting the seminary was given anew to the Lord and to his cause. "The Lord," writes Miss Fiske, "accepted the offering, but in so doing asked that they should give not only gold and silver, but one-half of the twelve teachers who were associated with her that year. Miss Lyon often said in after years, 'I little knew how much that prayer meeting would cost me.' She had not expected to be called upon to give up her own helpers."

Nor did she seem to be expecting calls for teachers from other lands. She was training her pupils to uproot selfishness from their hearts and lives, to be ready for any sacrifice, and so full of the missionary spirit that as teachers and mothers their lives should tell on the next generation. She was anxious about their spirit, but not about their place of labor, and never encouraged romantic ideas. "Most ladies," she said, "can do more for the missionary cause at home than abroad. Wives, mothers, and daughters have much to do to elevate the standard of liberality in those they love. Perhaps as daughters you should not be willing to have so much lavished upon you while there is so little given for the cause of Christ. By constant well-doing you may influence a brother or sister to consecrate all to him. You may even lead a brother to give himself to the missionary work."

These words spoken January 11, 1843, show that her object in founding the seminary was not, as has sometimes been stated, to train foreign missionaries. Yet because her training cultivated a Christ-like spirit, it prepared laborers for any field to which the Lord should call them. Her idea of missionary work included church work and philanthropic work of every kind.

A new era was about to open. Five days later Miss Lyon told the school, "I have said much to you this year on cherishing a missionary spirit, but very little on giving yourselves to the work abroad. We are now asked to furnish two teachers for the girls' school in Oroomiah, Persia, to go with Dr. and Mrs. Perkins, who are about to return to the work there." Requesting any who would like to consider the question to write her a note, she asked all to pray that the Lord would show whom he would send. Within an hour forty notes were in her hand; one of the shortest of them read:—

If counted worthy, I should be willing to go.

FIDELIA FISKE.

Miss Fiske was a niece of Rev. Pliny Fiske, whose departure for the Holy Land in 1819, when she was three years old, was among her earliest memories. It awakened in her an interest in missions that increased as she grew older. Before entering the seminary she had gained experience as a teacher, in Shelburne, Massachusetts, her native place. A graduate of the preceding summer, she had at once been invited to teach in the seminary and already her services were so valuable that it was a question whether she could be spared even to be a missionary. But this point was soon yielded. In the account Dr. Perkins gave of his visit to the seminary we find the paragraph: "Miss Lyon's description of Miss Fiske, whom she knew thoroughly and loved devotedly, was so commendatory, that I could then accept it only as the prompting of affectionate partiality for her pupil, though from the lips of a discriminating judge of character; but how soon I had ample reason to feel that the half, nay, the tithe, had not been told me. Miss Lyon and Miss Fiske were to select the second teacher from the many candidates after my departure. Miss Fiske, however, on laying the subject before her widowed mother, found her so strongly averse to giving up her daughter, that she was compelled to a negative decision." Another was selected in her stead. She, too, yielded to the unwillingness of friends and declined to go. On learning this decision, Miss Fiske was led to reconsider the question herself. After a sleepless night, she told Miss Lyon she was still willing to go, if her friends would consent. "If such are your feelings," said Miss Lyon, "we will go and see your mother and sisters;" and in an hour they were on their way for a drive of thirty miles through the snow-drifts. Being several times upset, it was nearly midnight before they reached the home on the Shelburne hills. It was Saturday, February 18th, ten days before Dr. Perkins was to sail. For a month the mother had thought the idea abandoned. But when roused from sleep by the unexpected arrival that wintry night, the errand was too well understood to need explanation, and little was said about it then. In the morning Miss Lyon said: "I came with your daughter because I thought I almost knew your feelings. I also give up a daughter. I have thought she might comfort you in your declining years and at the same time labor for our dear seminary with me till I go home. If we are to give her up, we shall, in so doing, understand as never before the gift of the Son of God." Before the Sabbath closed, the mother was able cheerfully to say, "Go, my child, go;" and other friends consented.

On Monday Miss Lyon returned to the seminary. "When she described to us the interview," writes one of the pupils, "her face shone like an angel's, it seemed to me, she was so joyful in the sacrifice." Miss Fiske followed on Thursday to find that the school had been occupying every leisure moment in the days between with sewing for her, and a very good outfit was in readiness. A farewell service in the seminary hall was conducted by Mr. Condit, and in the evening Miss Fiske met teachers and pupils for last words. Of that hour a pupil wrote: "Shall we ever forget how she implored us to live for Christ, how tenderly she entreated the impenitent, or the tones of her voice as she once more commended us to her God and our God? She wept not herself but smiled and said, 'When all life's work is done we shall meet again.' Tears and sobs were our only reply." At two o'clock Friday morning she was on her way to Boston, where the party embarked for Smyrna on Wednesday, March 1st. Miss Lyon, whose whole soul was in the matter, was with her day and night, seeing that everything was done for her which could be done. Of her own feelings she said: "I thought I knew something of self-denial in giving money, but I am thankful that I had something else to give, for there is an inner soul that was not reached before. If I have two idols they are the seminary and the missionary cause, and these were both God's before they were mine. It is easiest, safest, and sweetest to trust him." In later years she used to say, "When I have been most devoted to the Lord's work throughout the world, I have found that he was caring most tenderly for the seminary."

Instead of visiting Mrs. Banister or Mrs. Porter, as had been desired, Miss Lyon remained at Deacon Safford's for uninterrupted thought and prayer for her beloved charge. Returning to the seminary March 7th, just at the time of the afternoon exercise, she gave an account of the sailing, and closed with these words: "Young ladies, you have one less to labor for you, but I trust not one less to pray for you. The last word I said to Miss Fiske was, 'Pray, pray for us.' And as I watched her till she was lost to my sight I could but feel that with her last look on her native land she prayed for you. Will you not pray for yourselves?" She went from the seminary hall to meet her teachers. One of them wrote Miss Fiske: "Miss Lyon's heart was too full for us to ask her of your stay in Boston. She repeated to us her last request of you and asked of us, almost in agony, 'Is there one here to pray?' We went in silence to our rooms convinced that we had a work to do in our closets."

The next day Miss Lyon wrote to Mrs. Banister: "Our young ladies are more and more youthful every year, but they are so docile that ours is a very sweet home. There is more missionary interest than usual and more desire among some Christians to be prepared for the service of God. But alas! one thing is lacking —the mighty power of the Holy Spirit. According to former experience the harvest time for this year will be passed in four or five weeks. The short summer term is a favorable time for fixing impressions, but not for the work of conviction and conversion. Nearly sixty of our number are without hope. As teachers, as an institution, we greatly need a revival. I fear to go forward, I dare not stand still, I cannot go back. Will you not set apart a little time every day until you hear again, to pray that we may be taught what the Lord would have us do? For a few days I design to study two passages upon prayer—Luke xi. 5-13, and James i. 5-8; would you like to study these daily with me as you pray?"

At morning prayers, March 9th, she told the school why she had remained in Boston, and that she returned when she did, that she might find others to pray also, but now, as when she sought to find the missionary teacher, she could not ask any individual, lest she should not ask the right one. If any had a heart to pray she entreated her to do so. She dwelt upon the sorrows and prayers of Christ till Christians felt that they were almost strangers to sympathy with him in his sorrow for the world. The same day she wrote Mrs. Safford: "It is so seldom that I leave this beloved household, that my meeting with them after the absence of only a few days awakens tender emotions on both sides. Such things are trifles in themselves, but I believe the most trifling circumstances should be used for the same great end. With regard to efforts in behalf of the impenitent all is dark. But with a burden on my heart which I cannot describe, there is something in my soul like trust in God, which is like a peaceful river overflowing all its banks. Light can shine out of darkness, and I have great hope that we shall receive a blessing whether God shall permit Mr. Kirk to come to us or not. I have an increasing sense of the importance of a work of the Spirit which shall reach our whole number. You recollect Mr. Kirk's description of the difference between passing through the valley and rising up into a revival, and leaping immediately into one. We need experience of the first kind. On this account I query whether it may not be better to defer Mr. Kirk's visit. If he could stay two or three weeks I would as soon that he come to-day as at any time. But if he can stay only one, or even less, it is important that he come at the right time. His fear that he could not stay long enough is my fear. It seems to me desirable that certain difficult cases should come under the influence of a strong nature and warm heart like his, and we all need some stirring means. But my own will has been kept in an even balance concerning this. I am prepared to acquiesce in the will of the Lord. But let me beseech of you to offer every day a prayer on our behalf, till you hear again—which shall be soon."

The next letter Mrs. Safford received was from Miss Whitman. Miss Lyon had a severe cold and the physician feared lung fever. She was reserving all her strength for the school, and was giving connected instruction on the subject of prayer. One text was Abraham's prayer for Sodom, which she compared to a weight thrown into a balance that would have turned in favor of Sodom had ten righteous persons been found there. She was watching a trembling balance and entreating Christians to cast in their prayers. It had begun to turn when she was next able to write, March 17th.

"Beyond doubt the Spirit of God is moving on the face of the waters. My distressing doubt about using extra means has been somewhat removed. The interest in missions, in the general path of duty, and other like themes, seems changing to an increasing desire for the special influences of the Holy Spirit. I have been able to meet all my appointments, though sometimes I have concentrated into half an hour all the strength of three or four hours. I have had a short extra meeting for the impenitent every day. In these meetings I have no very definite plan, my waiting eyes are unto God. It is sweet to carry every burden and every care to him and from day to day the path has been made plain. I have no knowledge of future duty and I ask for none.

My lungs have not allowed me the privilege of individual conversation, but the teachers and others are instant in season and out of season."

"*March* 21.—A large number of conversions have occurred within three days. The Sabbath is of inestimable value. It is worth more than all other days in bringing thoughts into captivity to the will of Christ. In times of revival it seems to be the day he delights peculiarly to honor. The means we are using are so small and simple I can hardly tell what they are, and yet they are numerous. We simply walk by the light that is given day by day. Our regular business goes forward as usual, but we turn from other sources of social enjoyment or improvement, and gather up the fragments of time for seeking the divine blessing. The teachers are all of one mind and one heart, and are emphatically the leaders of the flock. I want you should pray fervently for ——. She retains her hope, but something in her revolts from everything social in feeling or action. I cannot find that an individual in the house has been able to approach her. I have met minds in a similar state, and have judged it best to avoid meeting her on the subject hitherto; but many things can be done in a time of revival which cannot at other times, and I hope I may have the privilege of doing something for her. It is not best she should know that her case is mentioned between us. We are witnessing interesting reconversions among those who have long called themselves Christians; but we have some cases that seem almost hopeless. They have passed seasons of conviction, perhaps indulged hope once or twice, and are now clothed in the self-righteousness of not being deceived this time."

"*March* 25.—Your letter gave me joy. I knew you were praying for us but I wanted to have you tell me so. We are in greater need of prayer than ever. Of the sixty over whom I wept and prayed so much when with you three weeks ago, only a remnant are now without hope. But some very trying cases are left."

"*March* 27.—We have set apart this day for fasting and prayer, to seek a blessing on our family, and on ourselves. It is a great thing to give up all business for a whole day that we may meet God in the inner sanctuary. I trust the day is brought as a willing offering, and that it will be accepted through the blood of the everlasting covenant."

"*April* 13.—In all my privileged experience with the work of the Spirit, this I think has been of unparalleled rapidity; and yet I have never witnessed more quietness, nor any less of reaction in the result. It has seemed like a sudden, heavy shower, but falling so gently that not a leaf or twig among the tender plants is disturbed; and then suddenly giving way to the beautiful sun and refreshing dews. Of sixty-six who came to us this year, strangers to Christ, all but six have accepted him as their Saviour. More than thirty in a single week. As teachers we have a great work in cherishing these tender plants. O, to follow Christ in this work! This desire rises from the depths of my soul with unwonted strength.. Shall we not have your prayers?"

A recent letter from a pupil of that year says that Miss Lyon could not rest till every wanderer was brought into the fold. For months she followed the last one continually with her prayers, till she, too, found the Saviour.

The following is from a letter to Miss Fiske, which was accompanied by the "Missionary Offering," a little book from which quotations have been given in preceding chapters. "You may ask how I found time to write it. The truth is my spirit was so stirred and my heart so burdened that I wrote without inquiring whether I had time or not. The scenes of the revival, the nearness of our next missionary subscription, the falling off of the receipts of the American Board, all combined to awaken unusual emotions. I was preparing a series of topics to present to the school, the substance of which you will find embodied in the book I began before the monthly concert in May. It was scarcely two days before most of the materials were gathered together. They soon assumed a tangible form, merely as a relief to my own feelings."

Each member of the school was presented with a copy of the book.

The fear of retrenchment in mission work talked of at that time led her to say to the young Christians before her, "How anxious I should be for you if you were now to be left without Christian watch, and how much more heathen converts must require it."

The influence of this revival in 1843 was lasting, and so also was the spiritual good wrought by Miss Fiske's departure. Teachers read with their sections Dr. Perkins's "Eight Years' Residence in Persia," and all missionary intelligence had new interest. Though it was many years before there were Woman's Boards to loan illustrative costumes, the record of a meeting July 2, 1843, says: "Just at the close, Miss Williams came in dressed in Turkish style, and seated herself in the Turkish attitude." At another meeting a pupil from Kailua enters in native attire, addresses another in Hawaiian, and together they sing in that tongue a verse or two, which all recognize from the tune, as the well-known hymn, "From Greenland's icy mountains."

Beside the regular missionary meetings which most of the school attended, a half hour or more was given in the sections every Saturday to a recitation on missions with the use of maps. They gave six or eight weeks in this way to the stations of the American Board. Attention was directed likewise to the work of other missionary societies in America and in England. The reading room was supplied with as many copies of the *Home Missionary* as of the *Missionary Herald* and they were in equal demand. The "Iowa Band" was just beginning its work. Much interest was felt in the labors of colporteurs among our German population, and in the efforts of the society for aiding Western colleges. Miss Lucy Lyon, who had charge of the senior class in these studies, wrote Miss Fiske, January 5, 1845: " I usually take two or three hours to prepare for this Saturday exercise, and am more interested in it than in any other recitation. The class seem interested also."

A journal letter to Miss Fiske was begun within a week after she left. The part first sent was five and a half months on the way. Her first monthly letter was received with great joy in four months from date. It was the beginning of a steady exchange between the seminary and her missionary daughters, which cements the bonds of faith and love, and stimulates prayer.

July 19th, a month after reaching Oroomiah, Miss Fiske wrote Miss Moore: "How often was I with you in spirit in my little state-room in the *Emma Isadora*. In that precious place I opened one by one those one hundred and thirty letters so kindly prepared for me to read on the voyage. There were many from those not Christians; some of these revealed feelings which I had never been able to elicit in conversation. There were many expressions like these: 'When your eye reads these lines on the broad waters, will you not offer one petition that I may not be lost forever?' 'Pray for me that my present feelings be not transient,—that I may come to Jesus now.' 'While you labor for Persia's daughters, will you not sometimes offer a petition for your unconverted friend on Christian ground?' With such requests before me, surely I was not wanting in subjects for prayer; and I used to try to pray for them day by day, but I was not prepared to hear, without deep emotion, what great blessings had been sent. Not a word has reached me yet from the seminary, but yesterday, on taking up the *New York Observer*, which the last messenger had brought, my eye fell upon the notice of a 'precious revival' in the seminary, in which it was said that 'all but six were rejoicing in hope.' Such unexpected intelligence overcame me. To the question, 'Why do you weep?' I could only point to the item in the paper. I need not say that in thought I have lived those weeks over with you. And who are those who still refuse to sit at the feet of the Lord Jesus? Are they some of my own precious section? Such a revival has brought my sister teachers deep interest and deep anxiety. I would have loved to share it with you. There is a sweet delight in pointing souls to the Lamb of God."

On the first Monday of January she wrote Miss Whitman: "In looking over Miss Lyon's suggestions for the observance of the day last year I cannot tell you how I felt as I read: 'Perhaps next new year's day will find some of you on a foreign shore. If so, we pledge you a remembrance within these consecrated walls.' I thought not then that privilege would be mine; but now I count your prayers the greatest favor you can confer."

To an associate at Seir she wrote: "Will not our God hear the prayers from a multitude of his people in our beloved land? If he does not, must we not feel that we are hindering the mercy drops all ready to fall?"

We can easily imagine the enthusiasm with which articles were prepared for filling a box for Miss Fiske, about a year from her leaving. It was not cooled in the least by the agreement that that offering should not diminish regular missionary subscriptions. No matter how busy the school life, willing fingers found minutes for work of that kind. Once it was by omitting the evergreen decorations of the seminary hall for thanksgiving evening, that the time might be given to work for a box which the village ladies were filling for Oodooville Seminary, Ceylon.

Rev. Dr. Hawes, who made the first annual address, made also the eighth. After referring to the progress which he observed in other respects, he noted the increase in numbers,—from four to forty-nine in the graduating class, from eighty to two hundred and forty-nine in the school; and reported the whole number of graduates as one hundred and five; of pupils, nearly one thousand. "A large proportion," he said, "are teaching; ten are giving their lives to the Indians in the far West, or to the heathen in the more distant East.

During the last year, over twelve hundred dollars have been contributed by teachers and pupils for the kingdom of Christ."

In the autumn of 1845 two of the teachers, Miss Reed and Miss Foote, left the country for India as Mrs. Howland and Mrs. Webb; another, Miss Porter, as Mrs. Pitkin, entered the home missionary field, and several of the pupils were about to engage in the same work. After mentioning these facts, one of the teachers wrote Miss Fiske: "Do you not think that Miss Lyon is full of joy? Her cup was almost full when, for two successive years, our contributions amounted to more than one thousand dollars, but how much more she rejoices to give her daughters to the work!" In a few months she was called to make an offering which touched her heart as no other had done, when three more teachers left the seminary for work abroad—Miss Moore, the associate principal, as Mrs. Burgess, with Miss Martha Chapin as Mrs. Hazen, for India; and Miss Lucy Lyon as Mrs. Lord, for China. Of these changes she wrote: "I have nothing to say but to ask that the will of the Lord be done, whether we are with or without means to carry out our plans. My only wish is for the furtherance of his kingdom. We know so little of the great plans of God that it is safest to leave all with him." But when she told the school of Miss Moore's plans, she could only compose herself by asking them to sing, "God moves in a mysterious way." It even told upon her health. Some months later she wrote Mrs. Burgess: "I suppose I am in danger,—my lungs especially. The excitement of giving up you and Lucy may have laid the foundation. But I am altogether reconciled to your going, and have not had the least regret, though my heart clings to you with increasing interest. Your first letter came Saturday. I have read it twice and lent it to Mr. Condit, though I expect to lend the original to but few, for I wish to keep it as long as I live. Your interest in remembering me is a great comfort, but not to you so important a duty as many that have a claim on you in your new home." To Mrs. Banister she wrote: "I have passed through many scenes of tender interest, concentrating the feelings of years into one, and manifestly increasing my gray hairs. I feel the loss of my two nieces—Mrs. Burgess and Mrs. Lord—socially, more than in our work, though both were very important to the school." The loss to the seminary was gain to the work in India, where Mrs. Burgess labored with the same faithful zeal till her death in 1853.

To Miss Whitman, the remaining associate principal, the prospect at the opening in 1846 looked dark. She not only missed the teachers on whom they had leaned, but found herself unable to take up all that Miss Lyon was forced to lay down. The latter was able to speak in the hall but a few times the first term. Once she said: "The effort of raising my voice warns me that it may continue but a little longer. If I say anything which will lead you to live as you would wish to live and to fit you for such an eternity as you would wish to enjoy, treasure it up. It often seems to me that I am doing for you my last work." In their need, Mr. Condit often conducted a religious service, as did also Mr. Hawks. Their labors were greatly blessed. There were ninety in the family without hope in Christ, and thirty of them had been under Miss Lyon's instruction for one year or more. The teacher who took Miss Lyon's place in meeting this class went from them one Sabbath evening feeling that she could not again undertake to hold the attention of ninety careless girls. The next Sabbath evening their meeting was the monthly concert, and on the second they had a preacher from abroad. Before she was called to meet the ninety again, sixty of them were rejoicing in Christ. Next to the preaching of the word, the most important agency employed by the Spirit in this work was the labors of a number of the middle class who banded together to help the teachers by their prayers and efforts. "That precious middle class" was Miss Lyon's name for them ever after.

In the following September Miss Lyon lost a long-tried friend in the death of Mr. Condit. He had been the pastor of the village church during all her residence in South Hadley, sympathizing with her in her trials, and sharing in her joys. September 10th she wrote Mrs. Burgess: "Our dear Mr. Condit is very near his home. The king of terrors is approaching with gentle step as if loth to take his prey. Here I am alone in this great building; no one near to interrupt my grief. I love this solitude, for tears and prayers in his behalf. The years of our acquaintance pass in rapid review. As I dwell on him as a friend, a Christian, a counselor, a pastor, sadness spreads over my soul. And yet it is not all sorrow. Heaven seems to be opening her gates to receive another servant of Christ."

He was preceded to the heavenly world by a young teacher from whom Miss Lyon had anticipated much, and whose loss she deeply felt. Referring to her death she said: "I thank God that I do not know that any pupil of this beloved seminary has died without hope in Christ. If there has been such a death, I have been spared the pain of hearing of it."

Not long after Dr. Perkins returned to Persia he wrote Miss Lyon: "We feel under great obligations to you for the deep interest you took and the efforts you made to secure for us such a helper as Miss Fiske. If under her fostering care a scion of Mount Holyoke Seminary shall spring up on the plains of Oroomiah to bless benighted Persia, I know you will feel amply rewarded. May the Lord give you the satisfaction of seeing Miss Fiske's mantle rest on many of your pupils."

Miss Lyon replied: "It is my opinion that the leadings of Providence justify our encouraging unmarried women to become foreign missionaries."

In June, 1847, as she was giving up another daughter, she wrote Miss Fiske: "I should love to tell you how the Lord has led me since we parted; how one comfort has been taken and another given, and how the promise, 'As thy days so shall thy strength be,' has never failed. And I should love to tell you how my heart goes with Miss Rice, as I send her forth like one of my own children. I commend her to your love, to your prayers, to a participation in all your labors, joys, and sorrows. May you both live long, abundant in labors, earnest in prayer, rich in faith, and at last receive an unfading crown of glory. Ask Mr. Perkins to be a father to another of my daughters."

Interest was added to the anniversary in 1847 by the reunion of the classes of '44 and '45. Nearly forty were present. The journal says: "Miss Lyon was delighted to welcome back so many, and invited the two classes to hold another reunion here ten years hence. After we returned from the church she stood in her accustomed place and gave us all a few parting words."

When Miss Lucy Lyon left for China, her successor in writing the "Journal for the missionaries who were once members of the seminary," was Miss Susan L. Tolman. After two years she too became one of its readers, as the wife of Rev. Cyrus T. Mills, in India. Eighteen copies were then needed to supply those abroad.

"We are accustomed to think," writes an early graduate, "that woman's missionary activity is of recent origin, but those of us who sat at the feet of Miss Lyon when she wrote the 'Missionary Offering,' and when she sent forth Miss Fiske and Mrs. Burgess, feel that she was in spirit the first president of the first missionary society, and perhaps the parent of all the others."

Yet it should be remembered that woman's gifts to missions began before those days. "Small as her pecuniary means are," said Dr. Anderson in 1839, "nearly one-half of all that has been raised in our country for publishing the gospel among the heathen, has been contributed by woman; and probably the number of active friends in that sex is two or three times as great as in the other."

The year 1847-8 was prosperous. The changes which Miss Moore's departure required had been made. The influence of the revival of 1846 was still felt in the consistent lives of the converts. March 9th Miss Whitman wrote: "The aspect of the special interest has been very different from that of last year. Instead of an overwhelming influence, the still, small voice seemed to speak to individual hearts, and so silent and unseen was the work that we never wished to speak of it as a revival, and there was no time when we did not feel the greatest solicitude lest the work should decline. Yet within a few weeks fifty have indulged the Christian hope." Miss Lyon said near the close of the year that the Holy Spirit had never before exercised his converting influences in the family for so long a time.

Her own health was so much improved that years of usefulness seemed to lie before her. The following summer she spent much time superintending workmen in making improvements in the domestic hall, and when the teachers returned in September, she told them that the new arrangements were then as perfect as she could make them. Committing housekeeping cares to others she turned more exclusively to the mental and moral interests of her pupils; and never did she manifest more enthusiasm in her plans, nor present truth with more unction than during this winter. She noticed this increased vigor, and said to one of her teachers, "I don't know why it is that my mind is so active. It sometimes seems to me that I am doing my last work."

She spent the winter vacation at Deacon Porter's, giving this reason: "I am borne down as never before with a sense of responsibility in teaching eternal truths, and I wanted to come to my 'resting home,' that in this quiet chamber I might seek for wisdom, grace, and strength for the great work. The teachers were urgent that I should go to New York to sit for a portrait. I was reluctant to decline their generous offer, but my picture seemed of so little consequence compared with a better preparation for my important duties that I could not go."

Formerly she had taken with her a long list of business items on which to consult Deacon Porter, but this time business was left behind. Mrs. Porter wrote of the visit: "Never since my first acquaintance with her have I elsewhere seen the principles of the gospel so strikingly exemplified as in her life. But on account of her active business habits and her constant planning for improvements in the seminary, I had not seen so much of the devotional state of mind as appeared in this visit. Her theme was Christ and the privilege of laboring for him and making sacrifices for his sake. I think I have never witnessed a nearer approach to the mercy-seat than in social prayer just before she left. It was almost the last sound of her voice I ever heard."

Early in the second term, Miss Lyon was suffering from a severe cold with nervous headache, when she became aware of a fatal turn in the illness of a pupil. Regardless of herself she went to the sick bed, spoke words of comfort, and bent over the sufferer to catch her replies. Her disease had progressed so rapidly that friends must be speedily summoned from a distance. It was the evening before the February fast, which had been anticipated with deep solicitude. With these anxieties upon her mind, Miss Lyon spent a sleepless night and on Thursday was able to say but a few words to the school. On Friday she met them both morning and afternoon and turned their thoughts to the celestial city, that as its gates opened to receive their companion they might catch a glimpse of its glories. With rapture she exclaimed, "O, if it were I, how happy I should be to go!" but added, "not that I would be unclothed while I can do anything for you, my dear children." To those out of Christ she said with great tenderness: "If one of you were on that sick bed, I could not take your hand and go down with you to the brink of the world of despair. It would be too painful for me. I should feel I must draw the veil and leave you." She urged them to come at once to Christ, not from fear of death, but because of his infinite perfections. In view of the tendency to excitement, she spoke upon passages which teach that anxiety for the future shows distrust of God, and said with great emphasis, "Shall we fear what he is about to do? There is nothing in the universe that I fear but that I shall not know all my duty, or shall fail to do it."

At family prayers that evening, she read the fifth of Second Corinthians, and went from table to go with the sorrowing father to the sick room, and see the daughter's happy look of recognition. She had feared that he would not arrive till too late for this and now was so grateful that she could not rest. After another sleepless night she met the school for prayer before the remains of their beloved companion were borne away. She read the hymn beginning "Why do we mourn departing friends." It was the last time her voice was heard in the school room. In the quiet secured to her that day, she slept, and at night appeared refreshed. But the mail brought word of the suicide of a nephew, without evidence that he was a Christian. This was an overwhelming blow and the night was one of anguish. On Monday it was evident that erysipelas had fastened upon her. Though its form was mild she realized that the danger was great, but said repeatedly, "The will of the Lord be done; I desire to be spared only to labor for him." At one time she tried to dictate some of her thoughts in regard to the school in case she should not recover, but was not able. At another, she said, "I should love to come back and watch over the seminary, but God will take care of it." By Thursday there was congestion of the brain. S. D. Brooks, M. D., was then physician both in the village and the seminary. He was untiring in his attentions, spending three consecutive nights in the sick-room. But the disease made rapid progress, and her lucid moments were few. Monday evening, March 5th, the voice of Dr. Laurie, her pastor, seemed to recall her to consciousness. He asked, "Is Christ precious?" Summoning all her energies to utter it, her last word was, "Yes." Noticing a continued effort, he said, "You need not speak; God can be glorified in silence." An indescribable smile was her reply. An hour later, God had taken her to himself.

The funeral was on Thursday. Till then the body rested in the little room opening southward from the seminary hall. Before the public service, the young ladies took their last look of the peaceful features. After prayer by Dr. Laurie, the three relatives present, the trustees, teachers, school, and other friends walked to the church. The journal says: "We were forcibly reminded of anniversary occasions, and the thought that we were following that dear form for the last time was almost overwhelming. Prayer was offered by Rev. Dr. Harris, of Conway, and by Rev. Mr. Swift, of Northampton. The sermon was by Rev. Dr. Humphrey, from the texts, 'The path of the just is as the shining light, that shineth more and more unto the perfect day,' and 'The memory of the just is blessed.' The hymns sung were those beginning, 'God moves in a mysterious way,' 'Servant of Christ, well done,' and 'Why do we mourn departing friends.' From church the procession moved to the grave, which is on a gentle eminence in the seminary grounds, a little to the east of the building. Gathering round it, the school sang, 'Sister, thou wast mild and lovely'—varying the hymn to suit the occasion, and were addressed in a few appropriate words by Dr. Laurie."

The lot, thirty feet square, is enclosed by an iron railing and covered by a growth of English ivy from slips sent by hundreds of loving pupils. A simple massive monument of marble bears on the west side the inscription:—

MARY LYON

The Founder of

Mount Holyoke Female Seminary

For twelve Years its Principal

and

A Teacher for thirty-five Years

and

Of more than three thousand pupils.

Born February 28, 1797.

Died March 5, 1849.

On the north side:—

"Give her of the fruit of her hands and let her own works praise her in the gates."

On the south side:—

"Servant of God, well done;
Rest from thy loved employ:
The battle fought, the victory won,
Enter thy Master's joy."

On the east side:—

"There is nothing in the universe that I fear but that I shall not know all my duty, or shall fail to do it."

MARY LYON'S GRAVE.

CHAPTER XII.

MISS CHAPIN'S ADMINISTRATION.

1849—1867.

THE day after the burial, the journalist wrote: "We need not tell you how sad our hearts were as we returned from that grave. Deacon Porter, who had been with us for several days, and the other trustees, comforted us by their sympathy and prayers; their wives joined the teachers in a prayer meeting last evening, and pledged us daily remembrance at the throne of grace. The trustees wish us to carry out Miss Lyon's plans as fully as possible, and we feel under sacred obligation to do so."

While Miss Lyon was planning the seminary she kept in mind the brevity of life, and wrote: "Much care will be taken to adopt permanent principles and mature a system which may outlive its founders. There should be such a natural division of labor that no department shall require persons of extraordinary ability. Superior gifts are very convenient, but they are rare, and any institution that can be carried on only by such persons would be likely to fall by its own weight. In the proposed seminary there could be no foundation for strength and perpetuity without such a system as could be conducted by persons of suitable qualifications, without extraordinary gifts." Accordingly she had simplified each department and reduced its details to such order that they could be definitely recorded. The results of experience gave these records increasing value for reference. Her associates had succeeded in carrying out her plans, but in their view, "superior gifts" would be required in her successor. It had often been said that if Miss Lyon should die, the seminary would either be closed or its character changed. To the teachers it seemed as if they could not go on without her. But they loved the seminary too well to desert it in its trying crisis. In a few weeks they were still further weakened by the sudden death of Miss Curtis. Miss Whitman, the associate principal, was taking a six months' vacation for her health, and was in Ohio when she heard of Miss Lyon's death. She set out at once, and was nine days on her way to the seminary, for she was far from being well. Next to Miss Lyon the teachers depended on her. She had entered the seminary in its first year; graduating in the second, she began to teach with the third, and was elected associate principal in the sixth. Since Miss Moore's departure she had been Miss Lyon's confidante. On the 18th of April the trustees elected her as principal and appointed as her associate, Miss Sophia D. Hazen, a graduate of 1841, who, after teaching at the seminary for seven years, had received Miss Lyon's promise of a year's leave of absence at the end of the eighth, but in the circumstances consented to accept the appointment for one year. Borne down with an almost crushing sense of the responsibility so unexpectedly thrown upon them, their dependence was in the promise of strength according to their day. And the Lord's strength was made perfect in weakness. Mr. Hawks came to their aid by frequently conducting religious exercises, while all other arrangements went on as Miss Lyon had planned them. And the Holy Spirit was present both with restraining and converting influences. Before the year closed, sixty had begun to rejoice in Christ, and a member of the graduating class wrote soon after, that she gave herself to him during the anniversary exercises in church. The blessings of that summer were a pledge of what the Lord purposed to do in the future.

It was soon plain that Miss Whitman's health was unequal to the work, and in the following spring she reluctantly sought release. Miss Hazen left at the end of the year, and not long after joined the Nestorian Mission in Persia, as the wife of Rev. David T. Stoddard.

In her letter of resignation Miss Whitman said: "The experiment of the year has been of great value, proving the excellence of Miss Lyon's plans and that they can be executed by ordinary minds. The teachers are pledged to carry out the principles of the seminary. The applications for the coming year are more numerous than last year, indeed the list is already full."

In the course of the summer over one hundred applications were refused, three hundred having been accepted, making an allowance for fifty failures. But failures were fewer than usual, and two hundred and eighty—a larger number than ever before—came at the opening of 1850-1. Examinations somewhat lessened the number. The spirit of other years was manifest on "moving day," when there were twice as many volunteers for the fourth story as could room there. The next year the school was larger still.

In 1852 one of the older teachers became the principal of a Holyoke offshoot at Tahlequah in the Cherokee Nation.

Some friends of the seminary thought it desirable to place a gentleman at its head. But those who knew it best deprecated such an innovation without greater apparent necessity.

On a cold October day in 1837 a good man from Connecticut sought for his young daughter admission to the new seminary. Climbing over the threshold without doorstep, they found Miss Lyon—in her plaid cloak and green calash—busy directing the workmen. As soon as she was free, she came and began at once to question the little candidate, who felt herself looked through and through. "We are full now," said she, rapidly rubbing her hands as she talked, "but if there should be a vacancy by and by we will send you word." In February word was sent, "If you can come at once there is a vacancy for you." It was Friday. On Tuesday she came. Almost the first pupil she met was Miss Whitman, whose kindness to the lonely stranger was the beginning of a life-long friendship between these future principals. The younger was Miss Mary W. Chapin. Her wise father was in no haste to have her graduate. He chose rather that she should add to her course the study of Latin, which she began a few months after entering. Miss Whitman's father was a man of similar breadth of view. Before her entrance his advice about continuing her studies was, "Do, my daughter, as a hundred years hence you will wish that you had done."

Miss Chapin had been connected with the seminary nearly every year from the first. Her executive abilities were seen by Miss Lyon, who knew at once on whom to depend for important service. Successful as a teacher in class work or in sections, she had also helped Miss Lyon so much in adjusting complicated arrangements that she became thoroughly familiar with the details of seminary business. When Miss Whitman and Miss Hazen left, the main responsibility fell on Miss Chapin. No one better understood the principles of the seminary or entered more fully into Miss Lyon's spirit. Whatever Miss Whitman or Miss Lyon may have said about the ability of "ordinary minds" "without extraordinary gifts" to carry on the Holyoke system, no one can deny that rare ability is requisite in the principal. The successors of Miss Lyon have been worthy of her. Delicacy toward the living forbids full utterance of the esteem that would otherwise be gladly expressed. In no case could the demand be greater than during the crisis of the first changes. In August, 1851, and again a twelvemonth later, the trustees recorded a vote of "thanks to the acting principal and teachers for the satisfactory manner in which they have conducted the institution the past year, and request them to carry it on the present year on the same principles." Though Miss Chapin was the real leader of that noble band, she shrank so much from responsibility that she would not even appear in the catalogue as acting principal. With a heroism that can be appreciated only by those who know the circumstances, she was caring for the seminary till another could be found. She was so sure that she could not take the place, and so confident that the trustees had the same idea, that she was taken by surprise when they told her November 18, 1852, of her appointment as principal. Quite overwhelmed she again assured them that it was impossible. But they left without further action.

At the same meeting they had made Miss Sophia Spofford, a graduate of 1846, associate principal. On the 23rd, Mr. Hawks announced to the school the two appointments. In the division of labor the morning exercises were taken by Miss Spofford, "on whom," the journal says, "it sometimes seems that Miss Lyon's mantle has fallen." Another glimpse of her is given by a pupil who writes:—

"The 4th of July, 1854, did not fall on recreation day, and because it was not thought best to use it as a holiday most of the girls agreed to wear some badge of mourning through the day. Many of those who did not care for the holiday were willing to do this, and when the bell called us to the hall for morning devotions, the desks, bell-ropes, and clock were draped in black, and a black knot was conspicuous on most left shoulders. No notice was taken of it then, but when we were gathered for the general exercise in the afternoon Miss Spofford said it was appropriate that these emblems of mourning should surround us on this anniversary,—there was occasion for mourning through all the land; and then she gave such a clear statement of the aims of the Kansas-Nebraska bill, and the dangers that threatened our whole country, as I believe few women of that day could have given. We were electrified. It was not only that she was so eloquent, but she had so ingeniously turned our 'jest to earnest' that she gave us the impression we were all emphasizing her thought."

In 1855 the state of Miss Spofford's health forced her to resign and her place was filled by the appointment of Miss Emily Jessup, who had been teaching in the seminary since her graduation in 1847.

The anniversary address in 1856, by Rev. Dr. Samuel W. Fisher of Cincinnati, Ohio, alludes to a Mount Holyoke Seminary opened the year before in Oxford, Ohio, and contains the following tribute to both institutions:—

"Mount Holyoke Seminary, revealing in a remarkable degree the practical genius of our country, has attained an influence superior to that of any other institution for woman's education in these United States; an influence not at all the result of accidental circumstances, but springing from the character of those it has sent forth. Its students, as teachers, wives, and mothers, are scattered from Maine to Texas, and everywhere are brilliant jewels that proclaim the richness of the mine whence they came.

"So highly do we value your policy, and so excellent have been its results, that we have transplanted it from the valley of the Connecticut to that of the Ohio. You generously sent us some of your most experienced teachers to assist us in our great work.

"It is a noble thing to have originated that discipline which has so distinguished this institution. But it was essential to its success, when transplanted to another soil, that you should send with it those who thoroughly understood it. We almost doubted whether in our free West, where youth knows little of domestic restraint, it was possible to realize the power of this original. That doubt has passed away. You have one daughter there. Yet it is not in one or two that you will see yourself reflected. Here and there through all this vast country other institutions like this will rise, and bless coming generations, until your daughters, bearing your lineaments, and breathing your spirit, shall vastly multiply your power for good.

"Nay, more, your scholars entering other institutions will so carry with them the thoroughness of mental discipline here attained, as to spread your influence far beyond seminaries peculiarly your own. For it is the attribute of a great soul to communicate its impulses even to minds little in sympathy with it; and it is the glory of a successful movement to impress its character upon, and quicken with its own life, systems unlike itself. And thus you have inaugurated a new era in woman's education; and even though your peculiar policy be not adopted, your influence will inevitably shape other institutions, and elevate throughout our land the standard of education for woman. And when that time shall come in which the actors of the present are weighed by deeds rather than words, it will be told that here in New England a modest woman originated a policy that spread itself abroad, everywhere elevating her sex, and through educated and pious mothers, strengthening the foundations of the greatest republic in the world. Long after the willow that weeps over her dust shall have decayed, the name of Mary Lyon will flourish among the daughters, not of New England alone, but of a whole great people."

Four years later, the seminary gave up other experienced teachers to open a similar institution in Painesville, Ohio.

In 1858, Miss Chapin was given a year for recuperation, Miss Julia M. Tolman having been appointed associate principal with Miss Jessup. Seven of the ten years since her graduation, Miss Tolman had taught in the seminary. In 1855 she became principal of the seminary in Willoughby, Ohio, whose building was burned in 1856. Though in frail health on returning to her Alma Mater, she entered with enthusiasm into the work. But to the great regret of all, failing strength compelled her resignation in 1860. For a time her health improved, and in 1862 she became the wife of Lucius A. Tolman, of West Roxbury, Massachusetts, whose decease in 1871 was soon followed by her own. Rev. Dr. Laurie, her last pastor, described her as "a lady of very quiet manners, thoroughly good, of a sweet disposition, very intelligent in her views of men and things, and a good counselor in practical affairs."

After five years of invaluable service Miss Jessup was given leave of absence for a year, at the end of which she tendered her resignation, but the trustees, earnestly hoping for the restoration of her health, refused to accept it till the end of another year. Even then she was not able to return. Since 1862 she has been teaching in the Western Seminary, Oxford, Ohio, a bright example of patience and usefulness even under the iron hand of disease.

Miss Tolman's successor in 1860 was Miss Catharine Hopkins, who became the only associate principal when Miss Jessup's resignation was accepted in 1862.

After the addition to the building in 1841, rooms on the first floor of the south wing were occupied by the family of Mr. Ira Hyde, who was employed to look after the premises, and purchase supplies for the seminary. In 1844 he was succeeded by Rev. Roswell Hawks. In 1855 Mr. Hawks removed to a house in the village and his rooms were taken by Mr. John H. Chapin, a brother of the principal, who during his stewardship of eleven years never seemed to exhaust his store of expedients for the comfort of the household.

From time to time improvements were made. The ravine between the monument and the seminary was filled up. Between two and three thousand trees were set out along the walks which lead down to the brook behind the seminary. A rustic foot-bridge gave access to the hill beyond. In 1852 pipes were laid from this stream to the main building and hot and cold water was introduced in every story.

The journal notes other matters as follows:—

"*July* 22, 1853. As we left the hall this afternoon, our cars were greeted with thrice repeated cheers by the workmen, in which we heartily joined when we learned that the last brick of the new wing was laid. This north wing is one hundred and twenty feet by forty, with four stories above the basement; in the latter is to be a chemical lecture room and laboratory and a large ironing room. About half the second story will be occupied by a lecture room for natural philosophy; a 'business room' and a parlor for the principal will be on the same floor; and at the end of the wing, wood rooms for each story, and an elevator for wood and baggage, as in the south wing. This addition will also contain fifty private rooms. We have now had about eight hundred applications for admission next year. In accepting candidates, particular reference has been had to advancement in Latin, in which the standard has been much raised during the last two years."

"*September* 29, 1853.—The school opens unusually full. Many of the entering class came early and have finished a part of their examinations before the term begins."

"*November* 30.—A few days before Thanksgiving some of our benevolent young ladies, observing the worn condition of the carpets in the large parlors, raised one hundred and sixty dollars among themselves in one day. Then two of them went to Springfield with Miss Chapin, and selected a carpet, with new furniture to match. Tongues now kept time with busy fingers till the last stitch was taken and the carpet ready for the floor. No less busy were tongues and fingers when portraits, windows, and pillars in the seminary hall and parlors were decorated for Thanksgiving, or when corresponding preparations were proceeding below stairs. Thursday was well filled. After the public service and the festivities, came preparations for the evening, when we met in the parlors about one hundred invited guests from the town. At eight o'clock we repaired to the seminary hall, where calisthenic classes entertained the company while refreshments were served. At the close of the evening, devotional exercises were conducted by our pastor, Rev. Mr. Swift."

"*December* 1.—Mr. Dickinson of Durham, Connecticut, formerly a missionary in Singapore, became interested in our seminary from reading the memoirs of Miss Lyon. He is particularly interested in astronomy. Collecting seven hundred dollars for the purpose, he purchased and gave us a telescope. Its object glass is six inches in diameter and its highest magnifying power four hundred. An observatory twelve feet high, with revolving roof, has been built for it a little north of the monument. The apple trees which would obstruct the view have been cut away. The telescope is now in its place and in use."

"*Fall term*, 1854.—Come and see the changes. In the seminary hall the dark settees are replaced by longer and lighter ones. Mahogany desks are on the platform. The walls, once white, are now light fawn color. Passing through the north door we find that the space-way leads straight to the north wing. On the right is a new reading room and a library."

"*July* 30, 1855.—The library room is finished. There are shelves from floor to ceiling on two sides; a gallery with iron railing gives access to the upper ones. On the other walls hang historical charts. Two long tables, covered with black velvet, stand near the center. About half the shelves are vacant notwithstanding fifteen hundred dollars' worth of new books. One thousand dollars' worth are soon to be added. The room is to be open half an hour every evening and each pupil may draw two books, to be kept not longer than two weeks."

The twenty-five hundred dollars just referred to were obtained mainly in Boston through the efforts of Deacon Safford and Dr. Kirk, one bookseller giving books to the value of five hundred dollars. The earliest mention found of books received was made May 3rd of the first year, when the trustees record a vote of "thanks to the officers and students of Amherst College for a gift of books, and to Prof. Hitchcock for a gift of minerals." Six years later they authorized an agent to solicit subscriptions to the amount of one thousand dollars for the enlargement of the library. In 1852 they added a small appropriation to a gift of the senior class of '51, but until 1855 the original reading room, twenty feet square, had served as library also. The number of volumes in 1857 was about three thousand. In 1864 it was voted to throw into one the library and reading room, and to open alcoves which friends of the seminary might fill, giving their names to them if they chose. Thanks were tendered to Dr. Kirk for proposing to fill one alcove. But before any steps were taken to execute this plan, the generous offer of Mrs. Durant in 1868, resulted in greater enlargement.

In 1856 the seminary was deeply afflicted in the loss of its faithful friend, Deacon Daniel Safford, who after long and severe suffering gently fell asleep in Jesus, February 3rd. His last labors for the seminary were in connection with the finishing of the library. His gifts to the first building fund amounted to $4,000, and at the time of his death he was the largest donor to the seminary. But his personal aid for twenty years—like that of Deacon Porter—had been far more valuable than money. The following extracts from a letter by Miss Catharine McKeen are quoted from his memoir:—

"There was always joy in the house when it was announced that Deacon Safford had come. Sweet are the memories of those evening gatherings when the teacher's day was done, and as children at home, we passed an hour with him. With a quick sympathy and delicate playfulness he inquired into the affairs of the house—wood, water, changes in the building, the wherewith by which the multitude should be fed, domestic and pecuniary matters generally. We brought our wants and perplexities and spread them out freely before him. But sweeter counsel did we take together on the spiritual welfare of our household. It was this which most deeply interested him, and the simple, earnest expressions of his own dependence on the quickening Spirit often brought us low with him before God."

"His benevolence was not shown in words alone; money and labor were freely given; and for none of his innumerable services or traveling expenses did he take the least remuneration. From its beginning he assumed the task of making for the seminary its large and frequent purchases of groceries and a variety of other articles. Our tables daily remind us of the generous donor of three hundred silver plated forks. Our library is a monument of his liberality and efforts. He planned and directed the introduction of water into different parts of the building, and repeatedly spent weeks in superintending building or other improvements. Even in his brief visits, he was busy looking about the house and grounds to see what was wanting. Still he never seemed to think that he did any great thing, and it was quite embarrassing to try to thank him for his kindnesses. Little did we think when he promised to bring his minister the next time he came, that when Dr. Kirk should come it would be alone, to speak in commemoration of our departed friend."

The anniversary of 1863 was honored by the presence of the governor, lieutenant-governor, and staff. They arrived early Thursday morning and were present at the closing examination and the reading of compositions. The latter were destined to accomplish more material good than usual. One was a plea for a gymnasium, earnest, sensible, convincing, by a member of the graduating class. The reading was scarcely finished when Governor Andrew was moved to start a subscription, and at the church three hours later he announced fifteen hundred dollars already subscribed, provided twice that sum should be raised. Before night the figures reached nineteen hundred dollars.

This joy was followed by sorrow before another night. Deacon Porter, who had given his seat in the crowded coach to a lady and was riding outside, was thrown to the ground by a sudden start, and fatal results were feared. But in December he was able to walk with crutches, and came with an architect to plan for the gymnasium. Mrs. Porter came also. Calling for the subscription paper, she wrote "Hannah Porter, $500," saying that it was a thank offering for her husband's recovery. As in each previous case of building, his visits were frequent till the work was done. The south wing was extended east as far as the north wing, and the new structure made to connect with each, thus enclosing a quadrangle. The gymnasium hall was eighty feet by thirty, and nineteen feet high, with trestle roof slightly arching. A gallery was added later. It was heated by steam and the same boiler was made to serve also a new laundry. But the war made it hard to raise funds. In March, 1865, a subscription paper stated that nearly four thousand dollars had been received, but that over twenty thousand dollars more was needed. The list was headed with the names of three of the trustees, A. W. Porter, Abner Kingman, and Samuel Williston, pledging one thousand dollars each, on condition that the twenty thousand should be raised during 1865. The trustees appropriated three hundred dollars for apparatus and the hall was ready for use in the summer of 1865, but the new part of the south wing had only floors and stairways until 1867. Meanwhile the slow-coming funds were needed more imperatively elsewhere. In 1865 a water tower in the rear of the main building, but communicating with each story, took all plumbing arrangements out of the building.

Miss Lyon's progressive plans for the course of study were gradually carried on. Advance in Latin has already been mentioned. In 1852 Miss Catharine Mc-Keen, a superior Latin teacher, was secured, who accomplished much in improving that department. Trigonometry was added to the required course in 1855, and history of literature three years later. Both had been taught before.

There were lecture courses in natural philosophy by Prof. Snell; in chemistry, by Prof. Chadbourne; in geology by Dr. Hitchcock; and occasional courses in architecture, history, botany, astronomy, natural history and elocution. In 1860, instruction in anatomy and physiology with use of skeleton and manikin, and lectures in hygiene began to be given by a physician resident in the seminary.

Under the admirable direction of Miss Eliza Wilder, who came in 1862, the standard of vocal music was much improved; the entertainments given by her classes on anniversary occasions included selections from the "Creation," the "Messiah," and other works of a high order. Private instruction began then to be provided for those who desired.

In 1861 a fourth year was added to the course, allowing more extended study in each department. Up to that time the entrance examinations had been oral. Since then most of them have been written.

For sixteen years the charge for board and tuition had been but sixty dollars for the school year. In 1854 the increased cost required an advance to sixty-eight dollars, and in 1857 to eighty; war prices in 1862 forced a rise to one hundred and twenty-five dollars.

It is sometimes said that school girls know more of the history of Greece and Rome than of their own country. However that may be in general, the daily throngs in the reading room, and the discussions everywhere going on, indicated intelligent interest in the war. At first all political parties were represented. In 1860 votes were cast and counted as follows: for Breckinridge, eight; Bell and Everett, thirteen; Douglas, thirty-two; Lincoln, two hundred and forty-six. A jubilant procession when Lincoln was elected and an illumination when he was re-elected were not the only indications of patriotism. Those who remained at the seminary in the vacation just after the attack on Fort Sumter, with the co-operation of Mr. Chapin, prepared a surprise for the returning family. It was the sight of the stars and stripes floating in ample folds one hundred and fifty feet above the seminary grounds. As everywhere else, hopes rose and fell as those colors hung aloft or at half mast. Needles were used for more practical purposes than flag-making. Within a week three hundred "comfort bags" were made and stocked with pins and buttons, threaded needles and other tokens of thoughtfulness. Into each bag went a little card with "Mount Holyoke Seminary" on one side and a verse from the Bible on the other. The letters from camp and hospital that came to "Mount Holyoke Seminary," laden with thanks from boys in blue, were a glad surprise. It was a delight to sew for them, and to knit army socks and mittens was better than fancy work. The spirit of all was voiced by one who was preparing the lecture room for the Soldiers' Aid Society, when she said she was glad to help, if it were only by bringing a chair; and what could be more appropriate than to spend Thanksgiving evening in the same work, foregoing for once the pleasure of extending customary hospitalities? And in similar employment sped the hours of Christmas eve. There were very few in the family who had not some near friend in the army and very few to whom there came no message of sorrow. In the prayer meeting for soldiers there were sometimes more requests than could be read. Here too working and praying were combined with giving, and the wants of the army were regarded as second only to missions. In 1864 the senior class gave to the United States Christian Commission the nearly two hundred dollars which they would otherwise have spent upon a class memento. At the suggestion of Mr. Mead, their pastor, then in the service of the Commission, that they should not graduate without a badge, Mr. Stuart of Philadelphia, Chairman of the Commission, ordered at the expense of himself and friends, the manufacture of scroll-shaped silver pins similar to those worn by the delegates of the Commission, bearing the class motto and name, and inscribed as follows:

The United States Christian Commission

To the Mount Holyoke Class of 1864.

"Sow we beside all waters."

Seminatores.

Only two or three in the audience knew of the surprise in store for the class when on presentation of the diplomas Mr. Mead called attention to the badge attached to the ribbon of each.

In May, 1862, invitations were sent to all graduates and many other friends, to join in a commemoration of the twenty-fifth anniversary, July 24. The number of graduates then living was eight hundred and sixty-four; one hundred had died. Three hundred and thirteen, including fifty-eight of the teachers, were present.

The seminary could entertain only about one hundred. The hospitable citizens of South Hadley cared for all the rest. Invitations to the collation on Wednesday were extended to all former pupils, and more than eight hundred were present.

There was an evening prayer meeting during the week in Miss Lyon's last room, and all the bustle of constant arrivals Tuesday night did not prevent a large attendance.

At nine o'clock on Wednesday, graduates gathered in class meetings for an hour and then—every class represented—took their places in the procession, under the direction of Mr. Byron Smith, who on that occasion took for the first time the place of his venerable father, E. T. Smith, Esq. The latter did not live to see another anniversary. Till that time with but one exception, the father had always been marshal of the day at anniversary, as the son has been since then.

Dr. Kirk, then president of the board of trustees, delivered the commemorative address, and afterward presided at the collation. The remainder of the afternoon was occupied with short addresses by Rev. Drs. Laurie, Hawes, Anderson, Hitchcock, and Rev. Messrs. Parish, I. P. Warren, E. Y. Swift, Daniel Tenney, and Mr. Samuel Burnham. At seven o'clock there was a reunion of graduates to hear reports from the different classes, after which the evening was given to a social gathering. A still larger number attended the usual anniversary exercises on the next day. Dr. Kirk's address had carried the thoughts of his audience backward twenty-five years. On this occasion Rev. Dr. R. S. Storrs carried them forward twenty-five years, predicting emancipation for the black man and "the perpetuity and prosperity of the American nation."

CHAPTER XIII.

MISS CHAPIN'S ADMINISTRATION—CONTINUED.

1849—1867.

IT was not the outward welfare of the seminary about which the teachers felt most solicitude, for without spiritual blessing no degree of success was counted prosperity. The care of souls outweighed all other care. While Miss Lyon was with them they had entered into her labors and shared her desires, but could not feel the same degree of individual obligation that they did when left without her. Instead of excusing herself from responsibility, each now took to herself a new share. Miss Whitman wrote June 1, 1849, of a visit from her brother, "He found that it is not one alone who is expected to take Miss Lyon's place, but that all the teachers as a band are endeavoring to do it."

Regular presentation of the truth, doctrinal and practical, Bible study in classes, studies on missions, and missionary meetings, all went on as before; personal effort was not remitted, but above all the teachers gave themselves to prayer Sometimes the answer came more speedily, and greater numbers were led to Christ than at other times; but as formerly, special blessings were given every year. At one time a teacher wrote: "We see again that while God works by means he does not depend on them. Whatever may be said about means and measures to promote revivals, I am more and more convinced that the most important means is prayer, and the most effective measure is prayer."

When Miss Lyon was in Boston at the time Miss Fiske left for Persia, she met at the house of Deacon Safford, his pastor, Dr. Kirk, who thenceforward took a warm interest in the seminary. His first visit was in 1844 when he made the anniversary address. He came next with Deacon and Mrs. Safford near the end of the summer term in 1855. The evening of their arrival he took charge of the weekly meeting. At the opening of the service, though uncertain in regard to the result, he invited any present who were seeking the Saviour and would like counsel, to come to him as to an elder brother. His discourse upon the parable of the prodigal son stirred many hearts. Christians retired to pray. More than twenty availed themselves of the opportunity for personal conversation, and at least one found peace in believing.

He spoke again in the morning, and by request of the school held a service the next evening. Many sought his personal counsels, and the work grew on his hands while he stayed. In two weeks he came again for a few days, and then for a few more at the end of the term. The Spirit accompanied his words with power. Perhaps none were made more impressive than these: "In all the future, let your religious duties be supreme. 'I must meet the Bible, I must meet the cross, I must meet Christ,' should be the language of every heart every day." In those few weeks there were about thirteen cases of hopeful conversion, but the work was mainly in stimulating Christians to a holier life. Before this no one had been called in for evangelistic labors, but Dr. Kirk regarded these results an answer to the question which he and Miss Lyon had often discussed—whether he could profitably labor with the school in times of religious interest. From this time he visited it at least once a year, frequently choosing the day of prayer for colleges. Teachers and pupils welcomed him as a friend and counselor. In 1856 he was elected trustee in the place made vacant by the death of Deacon Safford. From 1858 to the close of his life in 1874, he was president of the board.

"Yesterday, February 2, 1859," wrote a graduate of that year, "Mrs. Stoddard and her daughter Sarah from Persia, were visiting the seminary. As we were gathering about the tables for dinner, a lady whom we had never seen entered the hall with Miss Chapin. As soon as little Sarah caught sight of her she left her place and eagerly ran to greet her. Then we knew Miss Fiske had come. It is sixteen years this month since she left the seminary and Miss Chapin is the only one of our family whom she has seen before."

Miss Fiske's impressions of the seminary on her return are given in the following letter, dated February 8, 1859:—

"MY DEAR MISSIONARY SISTERS: Your journalist gives me space for a few lines, and I am glad thus to write to you all from our loved Holyoke home. I find it home here still and the spirit of 1843 unchanged. The whole teacher band and the pupils too have made me feel that I am not a stranger. It has done me good to hear these dear sisters pray for you. You may be greatly encouraged by this to labor in your blessed work; when weary, and when you lie down for the night's rest, you may be assured that the lamp still burns here. I find that sixteen years have greatly changed Holyoke, but I can rejoice in the changes, for they make our home a better one, and I love to think that dear Miss Lyon rejoices in them. Her spirit lingers here and her prayers are answered in the Lord's providing for every department. Miss Lyon's last charge to me was, 'Remember your duty to the seminary,' and I seem still to hear her saying so to all her missionary daughters. I trust that you are and will be more faithful than I have been to our Alma Mater. Write often, thus showing your undiminished interest in our old home, and more, tell your friends of your work, of your trials in it, that thus they may know how to carry you to their Father. Our work is one, whether at home or abroad, and it is deeply affecting to me to find so many here who love to make my cares and labors their own, and to give me their warmest sympathies and earnest prayers. There is far more of this than I ever dreamed of finding, and it makes me long to be once more in my Persian home, trying to serve my dear Saviour better than I have done."

At a later date she wrote: "To-day finds the seminary carrying out the principles on which it was established, as prosperously as when Miss Lyon left it, and with a larger number of pupils. It is no longer looked upon as an experiment and with suspicion; but it has a name and an important place among the educational institutions of our land."

Letters received from Miss Fiske and others who had gone from the seminary had contributed much to the interest of missionary meetings, and their frequent visits when in this country were more than welcome. Miss Fiske's return gave new impulse to this interest, and the months she spent at the seminary in the five subsequent years were perhaps no less useful than her sixteen years in Persia.

It was a happy thought that led Miss Chapin in 1859 to propose a reunion at the seminary of those Holyoke missionaries who were then in the country, inviting them to bring their husbands and children. It resulted in a representative gathering June 30th from many stations of the American Board, which was itself represented by Rev. Dr. Anderson. Of the fifteen ladies, five had been teachers here. Besides the forty-two from mission families, Mrs. Banister, Mrs. Cowles, and other friends made an assembly of more than one hundred guests. It seemed like a meeting of the Board in miniature. Short addresses were given in the seminary hall both morning and afternoon, interspersed with singing by the young ladies. The welcome was extended by Dr. Hitchcock, as one of the trustees. Dr. Anderson expressed his view of the importance of the seminary to missionary work by saying that to no college in the United States did he turn with so much interest while feeling the pulse of missions, as to this institution. He stated that sixty graduates had been missionaries of the Board and that twenty-eight were then in the field.

In the afternoon the missionaries occupied the platform, and were separately introduced to the audience by Dr. Anderson. As those from each station were named they rose to be recognized. Brief addresses from each gentleman followed, the tone of remark indicating the privilege felt in sacrifice for Christ's sake. One said, "It is worth all the suffering we experience to be drawn by it to Christ." It was ascertained that those present had performed an aggregate of three hundred and ninety-eight years of missionary service, besides Dr. Anderson's thirty-seven years as secretary.

The missionary interest that year was deepened by a visit of Miss Fiske in search of one to assist Miss Rice, whom she had left in her school with only native help. Several responded to her appeal and Miss Aura J. Beach of the senior class was the one appointed. Much of her outfit was prepared by the young ladies. Shortly before her embarkation in February, 1860, a party of forty went from the seminary to Amherst to attend the ordination of Rev. A. L. Thompson. That service was followed by a briefer one, in which Miss Esther E. Munsell, of Amherst—another Holyoke pupil—became Mrs. Thompson. Mr. and Mrs. Thompson were also to join the Nestorian mission, and to sail in the same vessel with Miss Beach.

Of some years not marked by special spiritual interest, facts like the following are given. In three weeks from the beginning of a year a section teacher writes: "One of the twenty in my care, and two others of the new scholars, have, we trust, accepted Christ. The change of rooms is to be made next week and many have already expressed a desire to have Christian roommates." One Sabbath five students agreed that each would pray specially for some one out of Christ. In two weeks they met again, bringing with them the objects of their prayers, all rejoicing in their Saviour; but instead of ten there were thirteen, for one of the number had a special interest for five persons and could not select one and leave the rest, and for four of them prayer had been heard. In two of those cases, as in others, the prayers of home friends had been unceasing. The mother of one wrote that the ladies of the church were praying for her and that her father felt that he could not continue preaching if his own children refused Christ. One wrote home at that time:—

"Lessons are learned and recited as usual, work is done and walks are taken; you may see two walking more slowly than usual and perhaps farther, that they may have more time to talk. From their tone and manner you may guess the nature of their theme; you might think it very still during prayers, but a stranger might stay for days in the family and never know how much feeling exists. Last Sabbath was a solemn day. Mr. Swift's morning sermon was on responsibility for the salvation of others, from the text, 'The voice of thy brother's blood crieth unto me from the ground.' Many were impressed with the truth enforced, that responsibility is not self-assumed and therefore cannot be laid aside at will. In the afternoon, the words, 'Come thou and all thy house into the ark,' were used to show the obligation resting on the impenitent. A few, we trust, are disposed to acknowledge this truth. Pray for them and for us all."

At another time there was unusual interest in the village church, which had set apart a day for prayer, and the pastor—Mr. Mead—proposed that instead of a forenoon meeting there should be a concert of prayer from closets all over the parish between the hours of nine and ten. "That morning," says another home letter, "Miss Hopkins found a note on her desk asking, 'May we not all be "with one accord in one place" at the same time?' In compliance, recitations were postponed for fifteen minutes that all who wished might assemble in the lecture room. The faith of those who thought only a few would attend was soon rebuked, for every seat was filled and extra chairs were placed in the aisles till there was no room for more. Afterward I heard one ask another, 'How did it seem to you?' 'As if Christ was there in person,' was the reply, and we all felt the same."

In July, 1860, the trustees expressed their appreciation of Miss Fiske's labors during the year and desired her to spend as much time in the seminary as was consistent with her other plans. She complied with the request and gave much of her time to the religious instruction of the school. Her apt and simple methods of explaining and applying Biblical truth interested and profited all who heard her.

The following account of the year 1862-3 is from the private journal of a teacher:—

The year was one of chastening. Of the three hundred pupils scarcely one had not a near friend in the army, and the silence that fell on the household at every allusion to the war showed the depth of feeling. Said Miss Fiske, who spent much of that year with us: "I can never speak of our country here without bringing tears to many eyes. It is heart work as well as work of the hands to which we are called." Tidings of sorrow came to many. In the first seven weeks five were called home by the illness or death of friends, and two heard of the death of soldier brothers. In the second term eighteen similar messages came within fourteen weeks; and during one week in July six were summoned to the bedside of friends not connected with the army.

One day a senior, scanning the papers for war news, came upon the list of killed at Port Hudson, and read in it the name of her own brother. She had no lesson the next morning. The father of another in the class and the brother of a third were wounded in the same engagement. A fourth had two brothers at Port Hudson and the home of a fifth was in danger from the invading army. Others were in suspense because no word came from father or brother. What wonder if that Butler lesson was hard to recite!

Throughout the year the daily or weekly prayer meeting for soldiers was well attended. In private interviews perhaps teachers prayed with pupils and pupils with each other oftener than had been their wont, but in other respects there was nothing marked. Out of one hundred not Christians when they came, twelve expressed hope in Christ during the first term, and on Sabbath evenings met in Miss Chapin's room for prayer. In the second term each week added to their number, but there was no general interest till the day of prayer for colleges. To that we had looked forward with unusual solicitude. On the Saturday preceding, a missionary address by Rev. Mr. Bushnell, of the Gaboon Mission, closed with an earnest personal appeal. At morning devotions on fast-day, Miss Hopkins read a note from him referring to the large company he had addressed on Saturday whose spiritual interests continued to weigh upon his heart. His message was as moving as his appeal had been, and his assurances of remembering us on that day reminded us of the multitudes who would also be praying for us. In appointing prayer meetings for the day, provision was made for notes of request. Perhaps a hundred were presented; one read like this, "Pray for me that I may be a better Christian." Others were as follows: "Please pray that one of your number who has long professed to be a Christian may lead a more holy life." "Pray for one in this family who has been following Jesus afar off." "Please pray earnestly for me that I may become a Christian to-day." Will you not pray for my roommate, that she may find her Saviour this very day." The meetings increased in size till the room could hold no more, and another was opened for the overflow. We thought of those who were praying for us and wished they could know that while they were yet speaking, God had heard. At the church service in the afternoon, the sermon by Dr. Eddy, of Northampton, was about the Syrophenician woman. It was listened to in almost breathless silence. In the evening each section teacher met the Christians under her charge. Miss H—— and Miss E—— met the others in the usual Thursday evening divisions, while the remaining teachers gathered to pray for all.

The next evening eighteen responded to an invitation given to those who had recently found Christ. The preaching of Mr. Mead, always so well adapted to our need, was peculiarly so on the Sabbath following, and in the evening the number in Miss Chapin's room was thirty. New voices were heard in recess meetings, some of them tremulous with confessions of returning wanderers. In one room the early part of fast-day found the occupants reading novels; in the evening they were praying that they might help each other live for Christ. In conversation with one young lady a teacher repeated the remark she had heard made about her, "Why! is —— a Christian? I never thought she was." Her distress that her light had been so dim led others to ask, "Can the same be said of me?"

March 9.—Interest in missions was deepened last October, when school was suspended for two days that all might attend the annual meeting of the American Board in Springfield. That we may become more intelligent in matters pertaining to the progress of the kingdom of Christ, a plan has recently been adopted for each section to appoint some of its members to collect geographical and historical facts about a given mission station, to report once or twice a week, once at the missionary recess meeting. Just now circumstances increase the previous interest. Last Saturday our hearts were moved for the needy children of New York City by an address from the superintendent of the Howard Mission. Yesterday a sermon by Rev. Dr. Hooker on Home Missions reminded us that the Howard Mission is not the only one in New York City, and that New York is not the only place where laborers are needed. To-day we feel not only that our country is in need but that the field is the world. At devotions this morning Miss Hopkins read a letter from Secretary Treat of the American Board asking for three teachers, one for the Gaboon Mission, one for Bulgarians at Eski Zagra, and one for Armenians at Marsovan. Some have been praying, "Show us an open door." Doors are open, near, and far away. In the course of the day, six went to talk with Miss Chapin or Miss Hopkins, others to other teachers, and more than half the senior class met for prayer in the evening in connection with the call.

March 16.—Six new ones attended the meeting for young Christians last evening. These met by themselves again to pray for those still without hope. Of that number seventeen went to-night to Miss Chapin's room, thus signifying their unwillingness to let this season pass by and leave them unsaved. This meeting was appointed with hesitation, for we are busy with our spring examinations; but it was thought that two or three might be benefited by it. Though the evening was full of extra engagements, when the bell rang, these seventeen did not forget their purpose. Miss Chapin says she never before knew it on this wise during examinations.

April 2. *State Fast.* — Prof. E. A. Park spoke at devotions on the fifty-first Psalm; at church on Galatians ii. 20. "I am crucified with Christ." The different prayer meetings were even more fully attended than on the last day of prayer, and overflow meetings went on at the same time. In one, three minutes were spent in silent prayer for those present without hope, a silence broken only by sobs. Of the requests presented, those for home churches and pastors were reserved till evening, when we assembled in the seminary hall. Some spoke of revivals already in progress; others were notes of thanks for answers to prayers previously asked. Before closing, Miss Hopkins called for one more prayer for those among us who do not pray for themselves. All Christian hearts were moved, and many had no voice to sing the hymn that followed, "Thou wouldst be saved—why not to-night?" We trust our prayers are already answered for one at least.

Friday, April 13.—Miss Chapin returned from Boston Saturday and at teachers' meeting Sabbath morning told us that she went away to confer with Mr. Treat, Miss Fiske, and others about the missionary teachers. Of those who responded to the call a few days ago, it is probable that two will go, Miss Fritcher and Miss Ballantine. To some of us this was entirely unexpected. The burden of the broken petitions that followed was that a spirit of entire consecration might be given to all of us.

At our monthly concert in the evening Miss Hopkins read letters from China and India, and at the close alluded to the call she had recently read to us, and said: "The one to go to Africa was needed very soon and before a response was received from us another had offered and been accepted; so that honor was not for us." But she added, "Our Father gives us the privilege of sending two of our teachers."

The breathless silence before their names were given, was followed by the sound of unrepressed emotion from Miss Fritcher's section, though, to prevent too sudden a surprise, she had told them of her plans before the meeting.

On Wednesday, April 8th, word came that the teachers must be ready to sail the last of May, too soon for Miss Ballantine to hear from her parents in India, and it is necessary to find one who can go without delay.

Summer term, May 7.—At breakfast Miss Chapin stated the position of the contending forces in Virginia. Our weekly meeting to-night was given to prayer for our army and country. The news of Hooker's retreat across the Rappahannock fills us with anxiety. We have many brothers there.

May 9.—Miss Hopkins read to us from the "Missionary Offering," and proposed that our annual contribution be taken up before Miss Fritcher comes to take leave of us and that so much of it as is needed be applied for her outfit.

May 21.—On Tuesday Miss Fiske gave some account of the Armenians to whom Miss Fritcher is going and of the nature of her work among them. She also told us of her own departure twenty years ago; of the willing hands that made ready her outfit in one short week and of her renewed courage years after, on taking up a spool of thread or a paper of pins and seeing "Mt. Hol. Sem." by the side of her own name. Then she named various articles needed by Miss Fritcher which any might contribute. Every suggestion was eagerly caught and Miss Fiske's room was soon crowded. Through the evening, packages kept coming till almost every useful article to be thought of,—from work box and writing desk with furnishing complete, to hammer and screw driver,—had been brought. And still the question came, "What can I give?" That none should be deprived, Miss Fiske gave another list on Wednesday, including saddle, riding suit, rubber suit, and other articles which could be procured by joint gifts. One thought of chintz for curtains and lounge covers. Another was sure there would be flowers to gather and brought a pair of vases. Anna B. said it seemed like the time in India two years ago when there was such a spirit of giving in the native churches.

Wednesday, May 27.—Miss Reynolds, of Springfield, who goes to Eski Zagra instead of Miss Ballantine, has made us a brief visit. For two weeks we have given all our spare time to Miss Fritcher's sewing, but to-night the lecture room is in order again, for the sewing is done, and the boxes are packed.

Thursday.—The work finished, we met in the hall last evening for a farewell service. In a few words Miss Fritcher expressed her thanks for the privilege of going, for the blessings with which she was sent forth, but most of all for the many assurances of continued prayer. Miss Hopkins repeated this assurance in behalf of us all, and once more we committed her and her work to Him who permits us to make this offering. Though there has been no formal laying on of hands we regard her as belonging henceforth to the daughters of Turkey.

From the hall the young ladies went to the parlors to say good-by. She left at five this morning. Most of the teachers rode with her to the railway station, but before taking our seats in the coach we went to number 111 for one more prayer in the place where we had so often met. Early as it was, a large number were waiting as we passed out of the gate and rode away. At Springfield she meets Miss Reynolds. Miss Hopkins and Miss Pond will be with them till they sail from New York, Saturday noon.

June 30.—This is Miss Chapin's birthday. We had planned a surprise for the evening, by assembling the school in the seminary hall arranged as a drawing room and then sending for her, after which strawberries and cream were to be served. Mr. Chapin was enlisted in the plan, the berries procured, and arrangements made in detail, including preparations for singing. But the news from the army and the prevailing anxiety took away all heart for festivity and the plan was abandoned. Miss Chapin and the school enjoyed the strawberries at supper quite ignorant of the purpose for which they were procured.

Saturday, July 4.—We celebrated the day by inviting the children of the Sabbath-school to spend the afternoon. The young ladies entertained them in the seminary hall, which was decorated as it was not on Tuesday evening. Though many older hearts were with the army at Gettysburg, to the children the world was all brightness. They sang songs, played in the yard, in the court, and in the hall. It was a pleasant sight at supper to see those fifty faces scattered about among us, some of them peering just above the table, others, though smaller, yet higher than they, because some big book made a common chair as good as the "high chair" at home.

July 16.—Rev. Daniel Bliss, of Syria, preached last Sabbath and spoke in the seminary hall in the evening. We have had visits also from Mr. and Mrs. Hazen, soon to return to Bombay, and from Dr. Gulick, who told us much about Micronesia. In response to an opportunity for asking questions nearly one hundred were written, embracing a great variety of subjects.

End of the term.—Within a few days two, we trust, have given themselves to the service of Christ. Nearly twenty are still undecided. It was sad to see so many go away without the love of Christ in their hearts. What a work it is to watch for souls! The sad and the joyful are strangely mingled in our thoughts, whether of school work or of our country. The news of **Vicksburg's** surrender and of successes in Virginia will always be associated with tidings of bereavement. In proposing a concert of prayer through vacation Miss Fiske reminded us that we must probably share still further in sacrifices for our country, and that though some have now no father, and not another brother to give, every one of us can and should give her country one faithful woman.

School opened September, 1863, with a family of three hundred and forty besides teachers, nearly two-thirds being old scholars, and the number without hope in Christ was even larger than the year before. Other cares so filled the time and thoughts of the teachers, as to take precedence of religious work. But Miss Fiske reminded them that they must not think too much of their own labors; if they were faithful in doing the work which God gave them, he would care for souls. They felt the rebuke when eighteen were found to have just begun to trust in Christ. By Thanksgiving that number was twenty-five. Early in February the teachers made special efforts to reach Christians personally. It was soon evident that the Spirit was deepening a sense of Christian responsibility in many hearts. February 4th, the teacher who conducted the weekly meeting with those out of Christ left it, overwhelmed with a sense of the powerlessness of mere human effort, and begged her friends to renew their prayers for divine power.

The following Sabbath evening, after meeting thirty of those not Christians, Miss Fiske said: "There was quiet solemnity in the meeting, but I sat with them in tears, for I found not one in earnest for eternal life. Many of them even refused to be spoken with."

On Tuesday evening, February 9th, after a sermon by Dr. Kirk on the parable of the rich man and Lazarus, fifteen remained for personal conversation, all but two professing Christians. Thursday morning his text was, "Strive to enter in at the strait gate." During recreation before supper the lecture room was crowded with those who met to pray; the hush at table was very marked. On leaving to meet an appointment in Albany, Dr. Kirk said, "It seems to me you are about to see a glorious work here. I wish I could stay and help you,"—and arranged to return Friday night. He was in doubt what to preach that evening and said before going to the hall that he felt as never before that he was only a messenger, and asked the teachers to pray for him,—not knowing that they had spent the preceding half hour in that way. He spoke from the words, "Choose ye this day whom ye will serve,"—setting forth the claims of God and pressing this point: "If you do not choose to serve God, you choose not to serve him—sitting here in these seats." He invited those who desired personal conversation to remain and asked Christians to retire and pray for them. The benediction was pronounced—but no one stirred, till a sign from Miss Chapin reminded Christians of his request. Teachers and pupils went out to give themselves to prayer, alone, or in groups. About forty remained in their seats in different parts of the hall. After speaking with each, Dr. Kirk asked those who wished to unite in a prayer of consecration to stand while he offered it, cautioning them not to rise unless ready to devote themselves to God. Some rose at once, others as the prayer proceeded. On one settee were three seniors; at first one stood alone, then the second rose, while the one between, who had for some time seemed to be almost persuaded, kept her seat to the end.

An hour after leaving the hall, the teachers were drawn one by one to Miss Chapin's parlor to learn about this after-meeting. Dr. Kirk could not speak of it without tears. As they told him of various perplexing cases he remarked that for forty years he had been studying to learn how to pray, and how to lead a soul to Christ, and found both more unfathomable every year. At breakfast Saturday, Miss Chapin found a note by her plate asking if the day might not be given to prayer. She replied by asking those who wished this each to write her a note to that effect. Within an hour, one hundred and forty notes came. School exercises were shortened and prayer meetings appointed as on fast-days. Whenever two or three found leisure moments, the time was given to intercession, and for a few days there were literally prayer meetings all the time. Miss Fiske said those days were like "the breakings down in the first years of the seminary, when the strong tide swept all before it."

On the Sabbath, Rev. Dr. Reed, of New Jersey, preached, and God used him to carry on his work. On Monday evening, February 15th, Miss Fiske appointed a meeting in the lecture room for those who had become Christians since the school year began. She wanted to see them in the same room where she met that company of one hundred and fourteen the first Sabbath of the year. She had a list of their names and asked those present to rise as she read their names. One rose—another—and another—then a name and no response. It was a solemn roll-call. Of the seventy-four present, thirty-seven had chosen the service of the Lord since the preceding Friday. So rapid had been the work that intimate friends recognized each other there with mutual surprise. Of the forty names to which there was no response, ten were at their homes and eighteen were at that hour in the north wing parlor as inquirers. Before the term closed, the seventy-four had become nearly ninety, and the number in the family bearing the name of Christ—about three hundred and twenty—was greater than ever before.

The joyful word was sent Miss Fritcher that all her section were in that number, and that those were praying for her who had never prayed with her.

The day of prayer for colleges found the teachers giving thanks for blessings received, and pleading for help in rightly guiding the young converts, whose future character would be largely determined by the first few months of their Christian life. One prayer meeting that day was devoted to the schools in Oxford, Painesville, and Oroomiah. The last was too far away to be heard from so soon, but before the term closed, from each of those in Ohio came news of revivals which began or received impetus on that day. In Oxford forty-one out of fifty-seven were rejoicing in Christ.

Just after that memorable Friday, allusion was made at devotions to the many letters written every recreation day, and to the influence that they might have when sent from such a place at such a time. Before the end of the term many interesting facts were reported in connection with the answers to those written after that suggestion. Pages might be filled with the answers to prayer. In March an opportunity was given for those who had presented requests, to write notes if they had reason to believe that prayer had been answered. The notes in response were more than could be read in half an hour. The experience of those weeks was to many like a fresh revelation from heaven that God hears prayer.

Almost every large ingathering of souls had been connected with a call for sacrifice. It was so now. In March Dr. Anderson came with the request for another teacher for Turkey. Though only one was called for, it was a matter, as Miss Chapin said, in which every one had a new call to entire consecration.

The close of the term is thus described by Miss Fiske, who seemed to have a special anointing for those last months of service.

"*May* 25, 1864. — I have been in my room much of the time the last weeks, but with care and rest have been able to attend meetings and point souls to Christ. Sometimes I have spoken to the whole school. Last Monday night I met the dear young disciples. They pledged to remember each other at the hour of sunset. Last night, even in the midst of carrying out the baggage, we had a prayer meeting of more than two hundred. It was one of the most delightful in which I ever had a part. A great company left about five this morning, but not till they had held a prayer meeting. When the second company of seventy were ready to leave we went to the library and had three prayers before we separated. They were all ready to go, and so we continued praying till the 'long bell' told us the carriages were at the door. It seemed so much like other years; just like that eastern home."

The summer term witnessed generous gifts and busy scenes like those of the year before, for Miss Pond, another teacher, was to follow Miss Fritcher to Turkey —making the fifth representative of the class of '57, in that field. Though she was to sail in the long vacation and final preparations could not be completed at the seminary, yet it was counted a privilege to aid in every way possible. But at this time Miss Fiske was not present. May 12th she wrote from her Shelburne home: "I cannot now make any effort without suffering, so I am very good in obeying medical advisers, and am really doing nothing. Those scenes of last winter were too much for me, but I shall love to go to heaven from them."

A shadow rested upon us which increased as the weeks went on and frequent tidings told us how she suffered. On Wednesday of anniversary week Dr. Kirk read to us at devotions her last message:—

To the Teachers and Pupils of Mount Holyoke Seminary:—

I cannot allow you to separate without blessing you once more in the name of the Lord. I want to thank you for those precious notes from you which have come flocking to my room for the past few days. I should love to write to you individually, but I cannot do it, nor even collectively except by the hand of another. Your notes have been an exceeding comfort to me, and your repeated assurances of remembrance in prayer have been more to me than any earthly good; let me thank you for it all, and assure you that Jesus will not forget it. I have loved you all tenderly and have loved to labor with you, and could I be with you this morning to give you one parting word it would not be a new one, but it is one which I would you should ever hold in remembrance—live for Christ. In so living we shall all be blessed in time and eternity.

Ever yours in the Lord,
FIDELIA FISKE.

A week later her sufferings were over. In the record book of the trustees, we find these words:—

Whereas, It has pleased the Great Head of the Church to remove from her earthly labors Miss Fidelia Fiske, once a pupil and teacher in this institution,

Resolved, That this board hereby express their appreciation of her eminent Christian character, her valuable services as a missionary, and especially her labors in this seminary after her return from the missionary field.

Resolved, That we regard the return of Miss Fiske to the seminary, her example, her religious instructions, and her entire personal influence in the seminary as the gracious reward to Miss Lyon for the sacrifice she made in yielding her to the missionary cause when her services seemed to be of great importance to the institution.

One of the last times Miss Fiske conducted devotions in the seminary hall, she spoke of the loss sustained in the recent death of Dr. Hitchcock, in whose wisdom Miss Lyon always confided, saying in every emergency, "I must consult my good friend Dr. Hitchcock." His last visit was at the twenty-fifth anniversary and his closing words to the school were those so often on his lips, "The everlasting foundations are sure."

The following minute was recorded by the trustees, July 20, 1864:—

Resolved, That, while in the death of Rev. Dr. Edward Hitchcock, the church of Christ has lost one of its most useful members, the ministry one of its ablest preachers, science one of its brightest lights, and education one of its most intelligent and influential leaders, Mount Holyoke Seminary mourns the loss of one of its fathers and founders, one with whom Mary Lyon took counsel in the earliest inception of her great plan, and to whom she and her successors in the government and instruction of the seminary have ever since looked up as a wise and faithful counselor. We are painfully conscious of the deep chasm thus opened in this board, which cannot be filled. At the same time we thank God that we have been so long permitted to enjoy his presence and to feel his influence; and we pray that a double portion of His Spirit may rest on us and all who shall come after us as the guardians of this sacred trust.

Readers of Dr. Hitchcock's works know that he attributed much of his success to his devoted wife, whose death occurred less than a year before his own. Friends of Miss Lyon from her girlhood, they had rejoiced together in all her success, and lived to see class after class take diplomas bearing the design first sketched by Mrs. Hitchcock's pencil. She was herself as a corner-stone "polished after the similitude of a palace."

Miss Chapin, in need of change, was voted leave of absence for the year 1864–5, and the care of the school devolved on Miss Hopkins, who had then been associate principal four years. Whenever Miss Fiske had been absent Miss Hopkins had conducted the religious services. These she continued in addition to the work laid down by Miss Chapin. Possessing rare qualifications she was further fitted for her position by ten years of experience in teaching. In the course of the fall and winter she had the joy of seeing forty out of sixty enter upon the service of Christ, and had looked forward eagerly to the last Thursday of February. With that day in mind, her subject at devotions on the preceding Tuesday was the unjust judge and the importunate widow. It was the last time she met the school. Inflammatory rheumatism from which she had suffered the preceding spring was returning. When Dr. Kirk bade her good-by Monday morning she was too helpless to give him either hand. Notwithstanding her sufferings she had continued during his stay to plan for meetings and arrange for those who wished to see him. In a few days the disease assumed a typhoid form and, as in the same month sixteen years before, delirium was followed by a sleep from which there was a glorious waking.

"Ye are not your own," was the motto she had given the senior class. "Freely ye have received, freely give," was the one Miss Lyon had given the class of 1849.

By the arrival of Miss Chapin and Mrs. Stoddard the teachers were not left alone in their time of need. But after having had the care of the seminary nearly fifteen years, Miss Chapin had previously decided to renew her resignation, which had not been accepted in July. At a meeting of the trustees March 29th, they passed the following resolutions:—

Resolved, That we fully appreciate the reasons assigned by Miss Chapin for relinquishing the position she has occupied so long, and the duties of which she has so ably discharged; and in accepting her resignation, we would express our high personal regard, our grateful appreciation of her valuable services, and cordial wishes for her usefulness and happiness in her new sphere of life.

Whereas, It has seemed good to God in his wise providence to remove from us March 10, 1865, Miss Catharine Hopkins, late associate principal :

Resolved, That as the board of trustees we hereby place on record our painful sense of the loss we have sustained in being deprived of her invaluable services at just this time when she was in the height of her usefulness as acting principal of the seminary, and when she was looked upon by all as peculiarly fitted to occupy the position of principal made vacant by the resignation of Miss Chapin.

Resolved, That we appreciate most heartily the prudence, skill, executive efficiency, and the conscientious faithfulness of Miss Hopkins during all her career as a teacher, more particularly since she has had the entire care of the seminary ; but we would express our special estimation of her single-hearted devotion to the spiritual welfare of her pupils, and of her able, clear, and impressive religious instructions.

The summer term began April 15th, a day never to be forgotten in our nation's history. To the flag, which had been arranged about the desks in the hall, to welcome back the household, a draping of black was added. The nation mourned the mighty fallen, and yet no loyal heart doubted the issue. So of the seminary; trustees and teachers had confidence in the principles on which it was established, faith in the promises fulfilled to its founders, and were firm in the belief that it had a yet greater work to do for the church and the world. Thrice bereft, without Miss Fiske, Miss Hopkins, or Miss Chapin, the teachers found in Mrs. Stoddard one who had not forgotten the feelings of those on whom the care devolved when Miss Lyon died. Complying with the desire of the trustees, she faithfully filled for two years the important position of acting principal.

Near the close of that time, Henry F. Durant, Esq., made the first of his many visits to the seminary. From 1863 the proportion of those who were Christians on entering steadily increased and the scenes of former years could not be looked for. But Mr. Durant's efforts to quicken conscience and waken a sense of responsibility were attended with the power of the Spirit.

May 17, 1865, a party of twelve teachers drove to Somers, Connecticut, and were witnesses of the ceremony that made Miss Chapin the wife of Claudius B. Pease, Esq.

In February, 1866, the young ladies rendered efficient aid in the outfit of another teacher for Turkey; this time a graduate of the preceding year, Miss Roseltha Norcross, of Templeton, the birthplace of Rev. Dr. Goodell. Her preparations were completed within three weeks from the time that she left her pupils in Templeton to go to Eski Zagra. She was to sail from New York the last Saturday of February with Miss Warfield, of Franklin, and Miss Seymour, of Rochester, N. Y.,—both under appointment for Harpoot. On the Thursday preceding the embarkation Dr. N. G. Clark brought Miss Warfield to the seminary to join Miss Norcross, and that she might meet the young ladies and gain an interest in their sympathy and prayers.

As the party took seats in the coach on leaving for New York they heard from the crowded balconies the hymn,

> Ye Christian heralds, go proclaim
> Salvation in Immanuel's name.

THE LIBRARY—NORTH SIDE.

CHAPTER XIV.

ADMINISTRATION OF MISS FRENCH.

1867—1872.

AT a special meeting of the trustees, June 27, 1867, Miss Helen M. French was elected principal, with Miss Mary Ellis and Miss Julia E. Ward as her associates.

Miss French was a graduate of '57, and save one or two brief absences, had been teaching at the seminary from that time. It was not without hesitation that she accepted a position so responsible and laborious, her health being never firm; but her two associates were specially commissioned to stay up her hands, and relieve her whenever she might be compelled to go away for rest. The journal, in alluding to her appointment, adds in Scripture phrase. "The hearts of the teachers do safely trust in her. . . . She is unquestionably called to this work, and with such helpers as Miss Ellis and Miss Ward, the burdens will not be overwhelming."

For two years, Miss French had conducted the morning devotions, and the trustees requested her to continue this service, leaving a part of the general business to her associates, both of whom had much experience. Miss Ellis was of the class of '55, while Miss Ward was a classmate of Miss French.

The term opened with a good attendance, and the newly finished extension of the south wing was occupied for the first time. This was destined to be an eventful year; and though not without its trials, it is memorable chiefly for its special blessings.

During the autumn, there began to be talk about the practicability of warming the building by steam.

Though the private apartments were comfortable, each with its coal stove, the long halls and public rooms were often cold. In winter weather, whenever one left her room there was a sudden transition to a low temperature, and exposure to chilling draughts sweeping through the halls. As the new gymnasium was warmed by steam, it was natural to think of advances in that line. The question was discussed at table and elsewhere with ever growing interest and hope. At first the trustees doubted whether in a building so extensive and so complicated as this had become, steam-heating would succeed, except perhaps in the public rooms and halls. Moreover, there was already a debt, and no spare funds. Yet since the need was pressing, and the courage of the ladies equal to a great effort in raising money, it was decided to look into the matter. The story of the enterprise is told in the following passages from private letters noting its daily progress:—

"*November* 20, 1867.—The engineers report that ten thousand dollars will cover the expense for putting steam in every room, and we teachers and scholars have resolved to try to raise the money ourselves. There is great enthusiasm on the subject. Hundreds of letters have been written within a week, asking aid from graduates and others. Most of us mean to see what we can do personally in vacation. We do not intend to refuse any sum, from ten cents to a thousand dollars! . . . Yesterday the mercury stood at fourteen degrees below zero."

The season continued unusually severe, and the day of return from vacation—December 11th—was bitterly cold. Every one was impatient to hear how much of the ten thousand had been secured.

"*Thursday, December* 12.—A teachers' meeting was called to-day to report success. Each stated the amount on hand, and also the amount definitely promised. Most of us had not yet heard from our letters. There was only one who had received nothing, and I came next. But no one knows how rich I feel with my silver quarter, the voluntary offering of a boy friend [silver money was a curiosity then], and my thirty unanswered letters. The amount in hand was over six hundred dollars; that promised, over three hundred, besides an indefinite amount *expected.* Miss French, with a graduate, gave two days to soliciting in Brattleboro', and received one hundred and eighty dollars. One man, passing over a two dollar bill, said he had not much to give, but he ' would not miss having a part in the prayers of those girls.' The mail this afternoon brought more money, so that the total received or promised to us teachers is nearly a thousand dollars. We took a similar report from the young ladies, and find that in all nearly two thousand has been received, and five hundred more pledged. . . . If our zeal wavers, we have only to pass a little way through these freezing halls to get fresh impetus."

"*December* 13.—Letters received by teachers to-day inclose eighty dollars."

"*December* 18.—The amounts pledged during vacation are now coming in. The mails for a few days have brought an average of one hundred dollars daily."

"*December* 19.—To-night just one hundred dollars. Last night it was only sixty-five."

One friend offered a barrel of apples instead of money. The gift was gladly accepted; and at a social gathering on the evening of New Year's day, the apples were retailed to a crowd of merry purchasers, clearing twenty dollars.

"*January* 15, 1868.—Last evening we received one hundred dollars from Mr. Durant,—the first gift above fifty, though a few hundred-dollar donations are promised. To-night came the largest amount yet brought by one mail—one hundred and seventy-one dollars."

The receipts varied greatly from day to day; but something for steam never failed to arrive, by every mail, for fifteen successive weeks. On Saturday of this week came two hundred and seventy-seven dollars; on Monday following only a dollar and a half. Every evening the teachers whose duty it was to record donations used to sit at the receipt of steam money. "Last night," says the record of January 30, "they waited and waited, but nobody came. Rather than close the book without an entry, they resolved to solicit something from the teachers, especially as the fourth thousand would be complete if they had only sixteen dollars more. While on this errand, news came of an overlooked letter containing twenty-five dollars." The winter term closed with a bank account of $4,680 for steam; and at anniversary, it amounted to $5,450.22.

Meanwhile another movement was in progress whose results were even more important. During the war, the enormous advance in prices, together with the cost of completing the gymnasium,—begun by contributions of friends,—and the purchase a little later of a lot on the north border of the grounds, had originated a debt, now over twenty-five thousand dollars. In 1864, aid had for the first time been asked from the state. The legislative committee reported to the senate April 15, 1864, their opinion "that the Mount Holyoke Seminary is eminently worthy of legislative recognition, and of the patronage of the state, in a grant of money from any funds that can be applied to educational purposes of this character; that in propriety and strength of claim it falls not below those institutions of learning which have been the recipients of material assistance from the state."* Yet the grant was not obtained, probably because of the demands for the war, and the uncertainty of public affairs. In 1866, Dr. Kirk renewed the effort, but, though the committee reported favorably, the bill did not pass.

At the session of 1868, the trustees asked for a grant of forty thousand dollars, having in view not only the

---

*The report instanced the grants in 1859, of $50,000 to Tufts College, and $25,000 each to Amherst and Williams Colleges and Wilbraham Academy; also $10,000 to the Massachusetts Agricultural College was voted in 1865.

liquidation of the debt, but also other needs. An able lawyer was employed to present the case before the legislative committees. The progress of the matter was watched with the deepest interest by the seminary family. On the twelfth of March, the newspapers stated that "the petitioners had leave to withdraw." But a few days later, Miss French was summoned to Boston to meet the committee on education. This looked hopeful. She was accompanied by Miss Evans,—who had been a leader in the effort for steam-heating,— and also by Deacon Porter, the treasurer. The committee made many inquiries regarding the special features and work of the institution. Their report shows that they were much impressed by several points thus brought out; among them, the inexpensiveness of education; the high standard of scholarship and character; the great number of teachers trained; and the value of the household work, not only in honoring labor, but also in forming habits of system, promptness, and self-help. A month later, the committee of finance made similar inquiries of Miss French and Miss Shattuck, with a like result. Both committees reported unanimously in favor of the grant. Even then, there were legislators who demurred at aiding institutions for women, but they were in the minority. On May 14th, the journal says, "At supper, Miss French read letters from friends in Boston announcing that the bill had passed." There was no doubt what Governor Bullock would do with it; and that evening there was a joyful illumination of the building from basement to cupola. The hundreds of happy girls who went out to see a sight so inspiring added their songs; and at length, gathering beneath the principal's window, they closed with the words, "Praise God, from whom all blessings flow."

The first thing that came of this forty thousand dollars was deliverance from the debt. From that day, the institution has been free. The wise management of its trustees and teachers keeps its ordinary expenses within its ordinary income;* and, unless in circumstances as exceptional as in this instance, it is not expected that another debt will arise. Next, it was made certain that the rest of the ten thousand for steam would not be lacking. And finally, there was the new library. It had been stated to the legislature that "a friend of the institution"— Mrs. Henry F. Durant— had offered ten thousand dollars for books, provided a suitable fire-proof building should be erected within three years. The appropriation made this possible; and thus secured a gift great in itself, and leading the way to other acquisitions of which at that time even the most sanguine hardly dared to dream.

At a special meeting of the trustees June 5th, the treasurer was authorized to receive the forty thousand dollars from the treasurer of the Commonwealth. It was voted to introduce steam, and to accept with thanks the generous offer of Mrs. Durant. Messrs. Durant, Kingman, Tyler, and Greene were appointed a committee on the plan, location, and estimates for the library building. Before the annual meeting in July, the contract had been made for the introduction of steam. The work was not entirely completed when the fall term opened, and the stoves still remained in the rooms; yet "it was a comfort," says the journal, "as we built our temporary fires, the day being chilly, to know that three and a half miles of steam-pipes had been put in during vacation." All was soon in working order. Coming back from a winter vacation was quite a different thing now. However cold the weather, one found a summer warmth pervading the building which made her quickly forget the chilly drive from the station. That first year, it is true, the steam had a way of announcing itself sometimes with a tremendous thumping and rattling, here and there, as if quite too sensible of its importance. Like a mischievous sprite,

---

*In July, 1867, the price per year for board and tuition was fixed at $150; fuel, lights, and lectures being additional. Since 1874, the charge has been $175, which covers the whole.

it would choose the most inopportune moments for its astonishing pranks. The next summer, however, some changes were made by which these annoyances were removed.

It was decided to erect the new library a little north of the main edifice, on the border of the lot purchased a few years before. The plans were drawn by Hammett Billings, afterward the architect of Wellesley College. The foundation was begun June 15, 1869.

At the next session of the legislature after the state grant was received, an act was passed, dated March 11, 1869, authorizing the trustees of Mount Holyoke Seminary "to hold real and personal estate to the amount of four hundred thousand dollars, in addition to the amount which they are now authorized to hold."

At the annual meeting of the trustees, the treasurer had the pleasure of announcing the receipt of four thousand dollars from four gentlemen, who shared equally in the gift. Thousand-dollar donations had hitherto been so rare that this was no less surprising than welcome.

At the same meeting, leave of absence for a year in Europe was granted to Miss Ellis. She went abroad in July, accompanied by Miss Shattuck, who returned in the autumn.

Toward the close of 1870, the library was completed. The exterior is unpretentious, yet tasteful and pleasing. It is forty-eight feet by thirty-three, with bay windows on the east and west. A corridor forty-five feet long leads from the main building to the entrance, which is protected by fire-proof doors. On entering, one finds a delightful room, well supplied with choice books; the ceiling is high, the well-lighted alcoves offer cozy nooks for readers, the furniture is of carved black walnut, and the floor inlaid. Steam-heated air is admitted from below. The cost was about eighteen thousand dollars.

In November, the books were removed to the new building, and the work of cataloguing began, the librarian—Miss Mary O. Nutting—having been appointed some months before. The method adopted was essentially that of the Boston Public Library. During the winter and spring, three thousand five hundred volumes previously in the library, and two thousand six hundred new ones, were catalogued. There were received from Mrs. Durant's donation about five thousand volumes in all; these were arriving from time to time during three or four years, mostly from Europe. Mr. Durant bestowed such care on their selection as only a genuine lover of good books could give. Occasional donations from other friends, and judicious purchases from year to year, have maintained a steady though moderate growth.

The new library at once became a constant resort. All day long, eager students were busy in consulting the authorities to which they had been referred. It furnished a perpetual stimulus to research. Everywhere the impulse was felt, and especially in the departments of history and literature.

More time was secured for English literature by requiring for admission some of the work previously taken in the first year. The study of ancient literature became henceforth the study, not so much of accounts of great works, as of the works themselves. Other tokens of progress deserve notice, including certain changes of text-books, and important additions to apparatus. The beginning of laboratory work by students is traced to this period. After a course of lectures on zoölogy in 1868, a natural history society was formed, including both teachers and pupils, whose zeal in collecting specimens deserves remembrance. Courses of lectures in astronomy from this time were regularly given by Prof. Charles A. Young, then of Dartmouth, who also lectured in physics. Prof. Charles H. Hitchcock, of Dartmouth, became the geological lecturer, thus continuing a work laid down by his honored father years before. Courses in elocution were given by Professor Churchill, of Andover, and Professor Bailey, of New Haven.

Early in 1868, some casual circumstance raised the question what proportion of all the graduates had deceased. Thirty classes had then completed the course. The memorandum catalogue, class letters, and other sources of information furnished the data; it was ascertained that the mortality for the whole period was between ten and eleven per cent. The triennial catalogues of colleges supplied the material for extending the investigation. In every case the period included was thirty years, ending about 1867. The result is shown in the following table:—

	No. Graduated.	Deceased.	Rate per cent.
Mt. Holyoke Seminary,	1,213	126	10.39
Amherst,	1,199	*135	11.26
Bowdoin,	1,012	*120	11.85
Dartmouth,	1,639	276	16.83
Harvard,	2,326	*268	11.52
Williams,	1,215	123	10.12
Yale,	2,883	387	13.42

*Exclusive of war mortality.

As these statistics were prepared so soon after the close of the period embraced, there were doubtless among graduates of each college a few instances of mortality not ascertained, which the next triennial included. The Holyoke percentage, as revised two or three years later, was 11.46, and that of Amherst 12.63. As the entire table was not revised, it is given as made out at first. It appeared at the time in newspaper articles by Dr. Nathan Allen, of Lowell, and Dr. Edward Hitchcock, of Amherst; and was afterwards incorporated in a paper prepared by request for Miss Brackett's book on "The Education of American Girls." It was also quoted in the *Westminster Review* of October, 1874.

"Our attention as teachers," says the journal in June, 1870, "has of late been called to the question, 'What are the peculiar features of Mount Holyoke Seminary?' The inquiry comes from Dr. Kirk and others who are trustees for the establishment of Wellesley College. Its founder proposes to make another Mount Holyoke, modified only in unessential particulars." The charter had already been obtained, and Mr. Durant was devoting much thought to the shaping of his plans. In regard to one of the points considered the journal remarks: "Perhaps we have all been surprised to find how much of the home-like character of the school, and the benevolent spirit as well as the aptness and efficiency of those trained here, may be traced to the domestic system. It was pleasant to have our views corroborated by the almost universal testimony of the senior class, each giving in writing her opinion as to the results."

In regard to the religious history of this period, there were many hopeful conversions every year, and evident spiritual growth of Christians. In October, 1867, a private letter says: "We have felt for some time that there was an under-current of special religious interest. One day Mr. Moody and Mr. Durant spent a few hours at the seminary; the latter spoke chiefly to the impenitent, the former to Christians. The hall was very still. We felt that the Holy Spirit was present in power. At the close, Mr. Durant invited all who were not Christians to meet in the south wing parlor, expressing the hope that they would go with the intention of consecrating themselves to the Lord. Of nearly fifty included in this class, thirty-two went. Much depth of feeling was manifest, and many offered prayer. Then they joined in repeating a prayer of consecration." That day and the days which followed, the hush of the Spirit's presence was felt through the family, and a new life was begun in many hearts.

The record of 1868-9 says: "There were at the beginning of the year forty-five not professing hope in Christ,—a smaller number than almost ever before. At the close, thirty-two of these classed themselves among Christians; they were led to the Saviour one by one."

In 1870, there was an interesting revival near the close of the winter term. It was preceded by weeks of earnest prayer. "We had had the promise of it," says the journal, "in the unusually fervent prayers of Christians, in the growing seriousness of the impenitent, in the thoughtful hush of the atmosphere; but it was on the day when multitudes throughout the Christian world were praying for us, that God's presence was manifest." Mr. Durant and Rev. H. M. Parsons, now of Toronto, conducted the public services and conversed privately with inquirers. "There was less opportunity than usual," continues the journal, "for prayer meetings; yet never was there more emphatically a day of prayer. Every one knew that the Holy Spirit was here. When one after another tremblingly or joyfully told of her new found love, we blessed the Master who had so abundantly fulfilled his promise. As was this day, so have been all the days since, except that we have attended to our usual school duties. The Lord has sent one and another to minister to us, each with a special message. . . . Among our two hundred and sixty pupils, only ten manifest no interest."

While Miss French was principal, four valued teachers left to engage in foreign missionary work. The first, Miss Olive L. Parmelee, of the class of '61, went in 1868 to teach at Mardin, Turkey. In 1869, Miss Elizabeth D. Ballantine, of the class of '57, married Rev. Charles Harding, of the Mahratta Mission, and returned to the land of her birth to carry on the work which her parents had left. In 1871, Miss Alice W. Gordon, a graduate of '67, and for two years a teacher at the seminary, went as the wife of Rev. William H. Gulick to a missionary work in Spain. Miss Fannie E. Washburn, of the class of '69, went in 1873 to be associated with Miss Fritcher in Marsovan, Turkey.

Near the close of 1869, the weekly Bible recitations, which had previously taken the place of morning devotions on Monday, were transferred to Sabbath afternoon at four o'clock. As attendance at church was no longer required in the afternoon, this change gave more time for the Bible classes. It soon became the custom at devotions on Monday morning, to devote the time for remarks to foreign missions, and other forms of Christian work. It has ever since been found helpful thus to glance at the great field, when about to begin the week-day routine.

It remains to speak briefly of some intimately connected with the seminary who in these years finished their course. Mrs. A. W. Porter died at Monson, December 16, 1869, after a long illness, aged seventy-two years. What she had been to Miss Lyon and to the seminary, in the early years, can never be fully told. She shared in the labors of seven revivals, and was always deeply interested in the spiritual prosperity of the school. She used to know the young ladies personally, and often visited them in their rooms. In her last illness, she desired the society of her seminary daughters. "Much of the time," says the journal, "we have gladly spared one, and often two, of our teachers to be with her. It was a sad pleasure to minister to her wants, and to treasure up in our hearts the lessons of love and patience taught by her sweet example." To the young ladies she sent as her last message, "Seek ye first the kingdom of God and his righteousness, and all these things shall be added unto you." "It was delightful," continues the journal, "to see that even when her mind wandered, her thoughts were in heaven. In the intense suffering of the last twenty-four hours, she several times repeated 'Let me go, for the day breaketh!' and once, in a transient respite from distress, her face brightened with a light borrowed from scenes invisible to other eyes, as she exclaimed, 'Beautiful! beautiful!' Her countenance—always delicate and refined—was very lovely in its last repose. A large group of her seminary daughters were among the mourners at her funeral."

A few months later occurred the death of another early and faithful friend, Rev. Roswell Hawks. An attack of paralysis closed his long life of eighty-two years, on Sunday, April 10, 1870, at the house of his brother, in Goshen, Massachusetts. It was fitting that, according to his own request, his grave should be in South Hadley. His body was brought to the seminary, attended by his daughter, Mrs. Putnam, and other relatives. The services at the church were conducted by the pastor, Rev. J. M. Greene. His text was, "In singleness of heart, fearing God," words aptly epitomizing the character and life of this friend. In the visit of Mr. Hawks to the seminary a few months before, it was beautiful to see that although his days of active labor were over, his devotion to Christ in all its fullness still remained. At the close of the service, the long procession followed the remains to the newly opened cemetery a quarter of a mile west of the church. The young ladies gathered around the grave and sang "Rest for the toiling hand," and then the venerable form was laid down to its peaceful repose.

At the next meeting of the trustees, they recorded the following tribute to the memory of Mr. Hawks and Mr. Southworth, another of the board, who died in December, 1869:—

It having pleased God to remove by death during the past year Rev. Roswell Hawks and Hon. Edward Southworth, members of this board, we cannot but put on record our high appreciation of the services they have rendered to Mount Holyoke Seminary, and the loss which it has sustained, as well as our own personal affliction in their death. Mr. Hawks was one of our most indefatigable collectors of funds in the early history of the seminary, and for many years the president of this board; he has never ceased to cherish a parental solicitude for its welfare. Mr. Southworth, during the fourteen years in which he has been a trustee, has been a constant attendant of the meetings of the board, a wise counselor, a liberal contributor to the funds of the seminary, and in every way a zealous promoter of its best interests as God has given him opportunity.

During the year 1870-1, ill health compelled Miss French to be absent, save occasional brief visits, and the associate principals were in charge. In July, 1871, finding herself still unable to resume her duties, she offered her resignation. The trustees declined to release her, still hoping that another year of rest would restore her strength,—a result earnestly desired by all. During most of the year 1871–2, Miss Ward also was obliged to be away, so that the administration devolved almost entirely upon Miss Ellis. Miss Spofford, a former associate principal, assisted her by conducting the morning devotions. It soon became evident that the health of Miss French, impaired not only by overwork, but also by repeated family bereavements, would never permit her to resume a position so laborious. In July, 1872, the board reluctantly accepted her resignation, and passed the following vote:—

Resolved, That we accept the resignation of Miss French with great regret, that we sympathize with her in the protracted ill health which has made it necessary, and that we hereby express our high appreciation of the wisdom with which she has presided over the seminary, and the gentle yet powerful influence toward Christian womanhood, which both by precept and example she has exerted over its pupils. Our best wishes follow her, for her future health and happiness.

Also the following tribute to Miss Ellis:—

Resolved, That we appreciate highly and gratefully the able and faithful services which Miss Ellis has rendered for fifteen years as a teacher in Mount Holyoke Seminary; and the very acceptable manner in which, in the absence of the principal and her other associate during a considerable portion of the last year, she has discharged the duties of acting principal; and we regret that these labors have so impaired her health as to necessitate her resignation.

THE OBSERVATORY.

CHAPTER XV.

1872—1883.

ADMINISTRATION OF MISS WARD.

AS the chairs of the principal and the first associate principal were thus left vacant, the trustees had now to elect their successors in office. Miss Ward was in Europe, and not yet able to return to her work. A month before, aware that new appointments must be made on account of Miss French's leaving, and that it might seem necessary for some one else to fill the position which she could not at once resume, she had sent her resignation to Dr. Kirk, at the same time expressing her hope that Miss Ellis would consent to become principal. Miss Ellis, however, had fully decided to leave;* and at the meeting of the trustees, July 4, 1872, Miss Julia E. Ward was appointed principal, her associates being Miss Elizabeth Blanchard and Miss Anna C. Edwards.

The telegram announcing these appointments found Miss Ward in Switzerland. It was arranged that she should remain abroad until October, her associates meanwhile carrying on the school. Both had been long connected with the institution, Miss Blanchard belonging to the class of '58, and Miss Edwards to that of '59. For two years, Miss Edwards had been principal of the Lake Erie Seminary, at Painesville, Ohio.

Various circumstances conspired to make this a critical period in the history of the seminary. Never

---

*After a few months of much needed rest, Miss Ellis went to spend some time in Europe. For a while she shared in the benevolent labors of Mrs. Dr. Gould at Rome. She was afterwards the lady principal of Iowa College.

before had educational matters held so prominent a place in the public view. Especially was this true in regard to the education of women. The time had gone by when it needed strenuous exertions in order to secure a single permanent institution of high order for girls; now, they were rising on every side. Every department of knowledge was making rapid progress; this naturally involved corresponding changes in methods of instruction. And thus, while the seminary had hitherto been constantly advancing, it was more than ever indispensable that it should advance now.

The newly appointed principal and her two associates were well fitted to meet the exigency which they so perfectly comprehended. They understood what the seminary was intended to be and to do; they believed in it with all their hearts; they devoted themselves to it without reserve. They were in the habit of thinking that what ought to be done could be done; and their courage and faith inspired those who shared their toils.

The school year opened auspiciously, under the management of Miss Blanchard and Miss Edwards. The late principal — now Mrs. Lemuel Gulliver, of Somerville, Massachusetts — kindly came for a few days to share in the labor of organizing. On the fifth of November, Miss Ward landed in New York, after a stormy voyage. Without allowing herself any interval of rest, she came to the seminary on the following day, to take up her work. It had been decided to provide increased facilities for the study of modern languages, without at present including them in the required course. French had been more or less pursued from the first; but hereafter there was to be a special teacher for modern languages. Miss Ward brought with her from France a lady of much ability and experience in teaching both French and German. Previous to her arrival, classes in both languages were in progress, instructed by a teacher who came several times a week from Springfield. During the year, nearly fifty students took French, and twenty-five German. Instruction in Greek also was provided, and several pursued the study of that language.

The chemical lectures this year were given for the first time by Prof. Charles O. Thompson, then at the head of the Worcester Free Institute of Technology. For the next ten years, he gave a course of about twenty lectures each winter. It was a matter of deep regret at the seminary when, in 1882, his acceptance of the presidency of the Rose Polytechnic Institute at Terre Haute made it impossible for him to continue services which were so highly prized. His sudden death, March 17, 1885, in the noontide of his usefulness, was deeply mourned wherever he was known.

For two or three years, there had been a feeling that a new building was needed, to be devoted to science and art. While its precise outlines and arrangements were not yet very clearly defined, every one felt that there were wants for which there seemed no other way to provide, and which a new building might be expected to meet. The telescope had been almost useless for some years, because the trees had overtopped the little building that sheltered it; there must be a new observatory. Former pupils, of whom not a few were in foreign lands, were often sending collections of plants, minerals, and other valuable objects, for the cabinets; and there was no room to bestow these fast-accumulating treasures where they could be accessible for study. A new building would not only meet the present want, but would doubtless attract many future gifts. And well filled cabinets were now quite as indispensable as books,—in scientific studies, even more so. An art gallery was desirable as another feature of the new building; for the general and growing interest in art, and the importance to a liberal education of some knowledge of its principles and history, could not be overlooked.

Thus the matter came to be frequently discussed among the teachers, and occasionally mentioned to friends. During the first year of Miss Ward's administration, a circular was issued, asking former pupils to send specimens for the natural history building which was beginning to be hoped for. Meanwhile the subject had been in the minds of the trustees; and at their meeting in July, 1873, it took definite form in a resolution to build, if the way should be clear to do so. This decision was greatly assisted by the generous offer of a member of the board, A. Lyman Williston, Esq., to give seven thousand five hundred dollars, one-fourth of the sum then supposed to be needful. The action of the trustees was publicly announced the next day, as well as the promised donation, though the giver's modesty did not permit the mention of his name. But the secret was easily guessed, and the pledge of so large a sum at the outset did much to encourage further donations.

That summer Professor Agassiz opened the Anderson school of natural history—the pioneer of the present host of summer schools—at the island of Penikese. Thither went from the seminary the enthusiastic teachers of botany and zoölogy, Miss Shattuck and Miss Bowen. Those weeks of eager listening while Agassiz, Guyot, and other great masters taught, and of ardent working under their guidance, were indeed fruitful and ever memorable to the crowd of teacher-students gathered there. Nobody could help imbibing a positive passion for scientific investigation, and for the laboratory work without which it could not be carried on. And thus the need of the increased facilities which the new building would afford, seemed to the Holyoke teachers still more imperative.

But there was much to be done before the way would be clear to build. The donation pledged by Mr. Williston, and afterwards increased to ten thousand, did not supersede the necessity of many lesser gifts, most of which must be laboriously gathered by personal solicitation. Moreover, in studying the various plans proposed, it became evident that there should be more ample provision for class rooms, cabinets, and laboratories; and although the thought of combining an astronomical observatory with these was given up as inexpedient, the estimates as finally adjusted amounted to fifty thousand dollars.

This, like the first of the seminary buildings, was destined to be erected "in troublous times." The financial panic of 1873 was just beginning. Men of wealth in many cases found their property so much depreciated in value that they hardly knew what they could call their own. It was about this time that a gentleman of property suddenly died, leaving a will in which was included, among other benevolent bequests, a large legacy to Mount Holyoke Seminary. But the will was not signed. Had the intended gift been received, it would have been an easy matter to build the new hall. It was doubtless wisely ordered, however, that this undertaking, like previous ones in the history of the seminary, should call for much exertion on the part of its friends, as well as constant reliance upon God.

Two recently chartered colleges for women were at this time building in Massachusetts,—one of them scarcely half a dozen miles from South Hadley. Times had changed within forty years, and each of these was to be erected by the munificence of a single individual. The founder of Wellesley College was a warmly interested trustee and benefactor of Mount Holyoke. Far from wishing to supersede or interfere with the work of the seminary, he proposed—as already stated—to reproduce its most important characteristics at Wellesley, and to attract thither from neighboring cities many who would otherwise be trained very differently. "There is no danger of having too many Mount Holyokes," he used to say with emphasis. Yet it was not very surprising that some should excuse themselves from aiding Mount Holyoke, on the ground that the new colleges would meet every want. "Let the seminary just go on as in the past," such would say; "if any of her students desire more than she can provide, let them go afterwards to Wellesley or Smith." Almost at the outset of the effort to raise funds for the new building, a city pastor, when consulted as to certain possible donors in his church, requested that they should not be approached. When it was urged that this was a critical time for the seminary, which was the pioneer and mother of schools for women, he replied, "The mothers die, and the daughters take their places."

Happily there were not a few who appreciated the need, and sympathized with the effort. To these it was clear that in the present circumstances, failing to advance would be virtually falling back. To cease growing would be to die. To take the position of a preparatory school, as some appeared to expect, would altogether subvert the aim with which the seminary was founded. Its training was meant to be preparatory to the work of life, and to that alone.

Dr. Kirk, the president of the board of trustees, was strong on this point. His successor, Dr. Tyler, of Amherst, remarks, "He always insisted on the necessity of keeping the seminary ever in the foremost rank of schools and colleges for the sex, as in Christian character and life, so in the standard of scientific attainment, literary culture, and all high and true womanhood. . . . While he contended earnestly for the faith and spirit of the founders, and insisted on keeping the seminary true to the principles on which, and the purposes for which, it was established, he was never afraid of innovations which were in the line of real progress. Indeed, he often remarked that the seminary was one great innovation, and Miss Lyon herself the greatest of innovators. Naturally, he felt the deepest interest in the erection of the new building for science and art."

In 1873-4, the seminary was crowded with students, more than three hundred being in attendance at the opening of the year. This was a larger number than ever before, excepting in 1863; it was in fact larger than could well be accommodated. In the circumstances, this was a cheering token; and others followed. Efforts to raise funds met with considerable success, hard as were the times. Now and then, gifts were arriving for the art gallery and cabinets that were to be. One of these, from the class of '71, was "The Morning Glory," an exquisite marble medallion on an easel, the work of the sculptor Jackson. During the two or three years while the art gallery was still waited for in faith, the "Morning Glory" sojourned in the parlor, the fair forerunner and pledge of the long train of beautiful gifts which have been coming ever since. The elegant engravings of celebrated paintings by Selous, "Jerusalem in her Grandeur," and "Jerusalem in her Fall," presented by the class of '61, came still earlier; and while they were temporarily adorning the seminary hall, silently prophesied to class after class, of beautiful things which they should by and by bring, as their own offerings.

The year 1874 is memorable for the departure of four whom the seminary counted among its chief friends. Mrs. Safford was the first to go. She had ably sustained Miss Lyon when few favored the enterprise, she had enlisted more than one of the seminary's firmest supporters, and she had never ceased to bear it upon her heart. In her last illness, her parting word for the pupils was, "Tell them to study the Bible till they love it better than any other book." To a friend she said earnestly, "Do you live much in heaven? Remember that One is interceding for you, that you may be prepared to enter there." Her death took place February 24, 1874. "She was conscious till the last two hours," wrote a friend, "and was full of faith, love, and joy in the Holy Ghost." Among her last expressions were, "Heaven grows bright! . . . I must die that I may live." Dr. Kirk, her pastor for more than thirty years, at her funeral remarked on the sacred words, "To die is gain."

Only four weeks had passed, when to Dr. Kirk himself there came the summons "to depart and be with Christ, which is far better." On Monday morning,

16

apparently in his usual health, he had attended the minister's meeting, and had spoken so ably on the topic of revivals, that his brethren requested him to continue the discussion the following week. On Friday morning, March 27th, while noting the outlines of his intended remarks, he laid down the pen and rose to take a few turns, as he was accustomed to do, in the parlor where he was writing. While walking to and fro, his feet faltered, his tongue and hands refused their wonted service, and he soon became insensible. A few hours passed, while tender and anxious ministries of friends and physicians surrounded him; then the heavenly gates were opened, and the blessed soul entered into eternal joy.

The tidings of Dr. Kirk's departure was a heavy and unexpected blow. Only two days before, as Miss Ward parted with him at his own house, he had promised her that he would soon come to the seminary for a visit. What his influence had been to both teachers and pupils can hardly be told. "The impression he made upon us in those days," says one of the class of 1859, "was that of a great and good man wholly intent upon drawing us away from that aimless, selfish life to which we were sure to be tempted, to one of earnest consecration to Christ." As the class of that year stood before him to receive their diplomas, he began his last address to them by saying tenderly and impressively: "Young ladies, your course of study here is now closed. It remains to be seen whether you will live for self, or for God." To the class of '71, he said: "You have selected a noble motto. Were you aware of its beautiful, comprehensive ambiguity? Was that your purpose? If so, I echo it in the fullness of its meaning.
'*Non nobis solum*'

breathes the humblest spirit of Christian dependence, the loftiest aspiration of Christian heroism. Not by our created strength and wisdom, not to our selfish earthly ends, will we live. This, dear young friends, disengages you from earth, identifies your will with God's, girds you with supernatural strength for every work and conflict, elevates your toil, purifies your affections, makes your earthly life celestial."

"Nothing was more manifest to all who knew him," says Dr. Tyler, "than his forgetfulness of self, and his entire absorption in the work of doing good to men and honoring the Master. . . . In childlike simplicity and humility, in spirituality and heavenly mindedness, and above all in prayerfulness, his prayers being the secret spring of his life, and his life being, in the language of Justin Martyr, 'all one great prayer,'— I think he surpassed all the men with whom I have ever had the happiness of being associated. There was the hiding of his power."

At the meeting of the trustees in July, Dr. Tyler was elected president of the board, which office he has never been permitted to lay down.

In December, within three days of each other, Mrs. Banister and Mrs. Eddy entered into rest. The death of the former, with whom Miss Lyon was associated at Derry and Ipswich, was to the Holyoke teachers and pupils a personal bereavement. She had been their guest only a few weeks before; and previous visits, particularly one a year earlier,* when she was prevailed upon to remain two months, had made her a dear friend whose counsel was much sought and greatly prized. Though now fourscore, the clear intellect, the rare conversational powers, the winning yet dignified manner, and the warm Christian benevolence that had marked her active years were still retained. In the course of

---

*In her life, lately published by the American Tract Society, there is an allusion to the seminary in a private letter:—

"Three or four times a week I have met the twenty-seven teachers and two hundred and seventy-five pupils in the hall. Miss Ward, the principal, is worthy of her position. Many of the teachers seem identified with the institution. Only one of them ever saw Miss Lyon. The wheels of this magnificent machine move on without much apparent friction; and the spirit of our Lord is manifest among teachers and taught."

these visits, she spoke to the school many times, and always with such aptness and ease that one would think she had never left the principal's chair.

Though never in firm health, and often a sufferer, she was habitually bright, cheerful, and full of interest in whatever was worth knowing or doing. At the time of her death she was spending a few weeks at the home of her step-daughter, Mrs. Hale, in Newburyport. It was preceded by only a brief illness, from neuralgia of the heart. Frequent paroxysms of mortal agony were patiently endured till the end came, December 3, 1874. "If the Lord will," were her last words, calmly uttered only a few moments before she ceased to breathe.

Of Mrs. Eddy, who was so closely connected with the early years of the school as Miss Lyon's associate and immediate successor, no fitting sketch can here be attempted. Though her health was never fully restored, the twenty-four years that passed after her leaving the seminary were tranquilly and usefully spent in her own pleasant home. Her love for the institution was ever warm, and for years her visits were frequent, and her counsels of great assistance. She died of pneumonia at her home in Fall River, Massachusetts, December 6, 1874, after an illness of only two days.

A somewhat full course of lectures on art was enjoyed in the latter part of 1874. The lectures were given by several of the Amherst faculty, and included short courses on architecture, sculpture, painting, poetry, and music. Not long afterwards, provision was made for studying the history of art in the senior year, as a part of the regular curriculum; and there have since been lectures in this department nearly every year. In French, German, and Greek, special courses were arranged, each occupying four half-years. While these are not intended to take the place of any part of the required curriculum, one or more of them makes a valuable addition to it.

On Sunday morning, January 18, 1875, it was discovered that the church was on fire. Already dense volumes of smoke covered the roof, in the southwest corner of which, around the chimney, the flames were just bursting forth. There was no fire engine, and not much water at hand, nor even snow. The building burned like tinder; in a few minutes the whole roof and spire were in a blaze. A few articles were snatched from within, and then those who were on the spot strained every nerve to save the dwellings near. Within an hour it was all over; the spire had fallen backward into the body of the church, which was already burnt bare—its "pleasant things laid waste." The beautiful new organ had been destroyed; the familiar seats, the pulpit, the communion table, hallowed by the sacred memories of so many years, were all gone.

In the afternoon a service was held at the seminary, and attended by many of the citizens. The pastor, Dr. Herrick, chose the text, "For my thoughts are not your thoughts, neither are your ways my ways, saith the Lord." For thirteen months, public worship was held in the seminary hall, which had been cordially offered by the trustees, and which by means of additional seats compactly arranged was made to accommodate between five and six hundred. At the close of the morning service, the Sabbath school followed; the evening meeting, however, was held in the small chapel, which had escaped the fire.

One of the pleasant memories associated with these months, when the church sojourned with the seminary, is thus related in the journal:—

"We are sure that few more interesting scenes can ever have been witnessed in the seminary hall, than the communion service last Sabbath afternoon (May 2nd), when twenty-five young persons made profession of their faith in Christ. Seven of these were members of our own family. Our pastor welcomed each with some brief and appropriate word of holy scripture, as he gave the right hand of Christian fellowship. The hymn was that one endeared to so many by its association with similar sacred occasions, "O happy day that fixed my choice."

As the seminary had owned a third of the church, a special meeting of the trustees was held February 10th, to consider certain questions in connection with the rebuilding. After conferring with a committee from the parish, the board proposed definite terms of agreement, which were accepted, and the new church was in a short time begun.

The site chosen for the new scientific building was a little northeast of the main edifice, with ample room on every side for the beautiful lawn and pleasant walks that were to surround it. By some happy fore-ordination, a magnificent black walnut tree, more than ninety feet in height, was already there; as if, scores of years before the building was dreamed of, it had been appointed to stand like a stately sentinel at its door.

The corner-stone was laid on the first of June, 1875. It was a bright and beautiful day. At four o'clock, the trustees and other friends, with the teachers and pupils, passed out through the pleasant grounds to the chosen spot. The outline of the future walls could be clearly traced, and it required no great effort of imagination to see the fine building that was to be. The company gathered around the southwestern corner of the foundation, and after Dr. Tyler had invoked the divine blessing, an original poem was read by one of the young ladies, and a paper relating the history of the enterprise thus far, by another. Dr. Seelye, of Amherst College, made appropriate remarks, in which he spoke of the present undertaking as continuing the work begun with so much faith and prayer by Mary Lyon. The corner-stone was lowered to its place by Mr. Williston. Underneath had been deposited a leaden box, containing catalogues and other documents relating to the history of the seminary, with stereoscopic views of the buildings and grounds. The exercises were closed with prayer by Dr. Herrick, the pastor, and the doxology.

LYMAN WILLISTON HALL.
MOUNT HOLYOKE SEMINARY — SOUTH HADLEY, MASS.
(FOR SCIENCE AND ART)

Notwithstanding the financial pressure, efforts to raise money were continued with some success. During the term, the principal devoted much time to this work, and obtained several thousand dollars; other teachers also labored effectively for the same object. The amount secured was made up chiefly of comparatively small sums, and represented many donors. Sometimes between two gifts came many unsuccessful solicitations. One day, eighteen gentlemen were visited without a single contribution being received, although expressions of interest were numerous and evidently sincere.

As the church was not finished, the anniversary exercises of this year were held in a tent spread in the grounds east of the library. "It was such a novelty," says the journal, "to see our white-robed procession turning into the grassy field, and after a little winding among the scattered apple trees to find ourselves in the cool, airy tent. We sat facing the south, with the library on the right, and the rising walls of our new building some distance to the left. The weather was deliciously cool, our speaker, Dr. Seelye, of Amherst, seemed in his happiest mood, and the address, delivered without a scrap of a note, was beyond praise. Everything was so delightful that we wished we could have our exercises in a tent every year."

Much was done in the line of internal improvements during the summer of 1875. Steam was introduced for cooking, in connection with the enlargement of the heating apparatus to warm the new hall; and many other changes were made in the domestic department to lessen labor and save time. At this time also, rowing was first provided for. Three beautiful boats were presented for the use of the young ladies; a boat house has since been built, and other boats added. Ever since, rowing has been a favorite diversion.

A pleasant incident of the next term was a meeting at the seminary of the Connecticut Valley Botanical Society, of which Miss Shattuck was a prominent member, and at one time, president. "You will recollect that we have had the honor of entertaining this highly agreeable body once before," says the journal. "On this occasion we had thirty or forty guests, including the distinguished Professor Gray, of Harvard, Professor Blanpied, of Dartmouth, President Clark, of the Massachusetts Agricultural College, and many other gentlemen as well as ladies, from New Hampshire, Vermont, and Massachusetts. At the afternoon session, as many of us as could find room in the library were admitted to hear the discussions, and found them no less entertaining than instructive."

Early in 1876, by request of the Commissioner of Education, a historical sketch of the seminary was prepared by Miss Nutting. An edition of this was published at Washington, and sent by the Bureau of Education to many institutions from which similar sketches were desired, "as a specimen of the work in preparation for the Centennial of 1876, and as covering the leading points of inquiry." A large edition was published by the seminary, and widely circulated in connection with its exhibit at the Centennial. The sketch was found useful also as a reply to inquiries regarding the history and work of the institution, which are always numerous. In 1878, another edition was required, a considerable portion of which was used at the Paris Exposition.

In April, 1876, another fire swept away the hotel, stores, and post office, with several other buildings; and for a time it threatened to destroy many more. The flames spread with fearful rapidity; there was no fire-engine nearer than Holyoke, nor any way to summon one save by sending a messenger. Fortunately the evening was mild and still. The seminary family promptly responded to the appeal for their personal help, and a "bucket line" was quickly formed,* conveying

---

* The seminary has since been amply supplied with hose and hand-grenades, in addition to the fire-extinguishers and force-pumps then on hand. There is also a fire-escape.

water from the nearest faucets, through the front door, and over thirty rods along the street. Meanwhile the steam pump was set at work to keep the cisterns full, and some of the young ladies were carrying water from private houses near the fire. People whose homes were in immediate danger were bringing valuables to the seminary, and numbers were beginning to come in from the neighboring villages. The roads were in such a state that it was an hour and a half after the alarm was given before the engine arrived. The progress of the fire having been delayed just then by the distance of the next house, the firemen soon extinguished it. Had it not been stopped at that point, it would doubtless have destroyed all the remaining buildings on that side of the street, and perhaps the seminary also. The village people were disposed to give a good deal of credit to the ready and resolute assistance of the hundreds of young women. The burnt portion of the street was soon rebuilt, partly in brick, so that the appearance of the village was changed only for the better.

A course of lectures on "Science and Religion," by the Rev. Joseph Cook, in the spring of 1876, awakened great interest at the seminary. Among the subjects were, "The Certainties of Religious Truth," "Causes of Skepticism in New England," "Decline of Rationalism in Germany," "Evolution," and "Materialism." Numerous question on the themes so ably presented were brought by the students. These questions, after having been somewhat classified, were answered by the lecturer, at two or three evening gatherings of intense interest, which were additional to the lectures.

CHAPTER XVI.

ADMINISTRATION OF MISS WARD—CONTINUED.

1872-1883.

IT was expected that everything would be ready for the dedication of the new hall by the middle of October, 1876. During the vacation a tasteful iron fence had been put up along the entire front of the grounds. Around the new building, walks had been laid out and shrubbery planted. The work of removing the contents of the old cabinets, unpacking specimens lately received, and duly arranging the whole, proved so arduous that another month was required. From the opening of the term, however, recitations were held in some of the new class rooms; and the first course of lectures was given in October, by Professor Hitchcock, of Dartmouth. Meanwhile the abundant labors of Miss Edwards among the minerals, of Miss Clapp in the zoölogical cabinets, and of other teachers in other departments, began to bring order out of chaos. As there had not hitherto been room to collect and arrange the treasures accumulated within the last two or three years, all were surprised at the result.

The dedication of the Lyman Williston Hall—as the new building was henceforth called—took place on Wednesday afternoon, November 15th. The occasion was one of hearty congratulation and devout thanksgiving. As the principal afterwards said, " When this long-desired building stood before our eyes complete and beautiful, it seemed as truly a gift of God as if it had come down from heaven."

An hour before the public exercises, the trustees met in their beautiful room, to receive the report of the building committee, and to make suitable acknowledgments of services and donations. It appeared that the total cost of the building and furniture, with that of grading and ornamenting the grounds, amounted to $50,017.74. To cover this, there had been contributed $30,914.22, and the balance of $19,103.52 had been advanced from the treasury; so that not a dollar remained unpaid.

At three o'clock, the trustees joined the seminary family and guests in the art gallery. After singing, the audience listened to the statement of the building committee, presented by Hon. Edmund H. Sawyer, and also to resolutions of thanks passed by the board. The address was delivered by Dr. Tyler, whose theme was the value of scientific and art studies, and the appropriateness of dedicating to God the building erected for them. Before closing, he paid a hearty tribute to the friend "whose name, in spite of himself, trustees, teachers, and pupils have spontaneously, irresistibly, and unalterably fixed upon the edifice." The generosity of other donors was gracefully recognized, as well as the persevering labors of teachers and friends in soliciting funds; particularly of "one to whose large plans and far seeing thoughts we are much indebted for the idea of this building as we are to her steadfast and unyielding purpose for its execution, and who has shown the same wisdom and skill in raising money which she has long exhibited in teaching and presiding over this institution. And she has found willing and efficient helpers in those who are associated with her in the government and instruction."

It was estimated that as many as five or six hundred persons had contributed either money to the building fund, or specimens to the collections. One gift of rare beauty and value was placed in the art gallery the day before the dedication of the hall; a four-thousand-dollar painting by Bierstadt, of "The Hetch-Hetchie Cañon." It was presented by Mrs. Edmund H. Sawyer and Mrs. A. Lyman Williston; the artist, himself a friend of the seminary, sharing generously in the gift.

The Lyman Williston Hall is a tasteful and substantial edifice, sixty-six feet by sixty-three, with a wing forty by twenty-four. The interior finish and the furniture are of ash. Like the other buildings, it is heated by steam. In the basement, a room sixty-three feet by thirty-two is devoted to the ichnological collections; and in the basement of the wing are the chemical storerooms. On the first floor are the trustees' room, the physical laboratory, the botanical and zoölogical rooms, together with a large lecture room, amply provided with every convenience. The chemical laboratory, in the wing, opens from the lecture room. On the second floor are the cabinets of minerals and of zoölogical specimens, the geological class room, and a complete set of Ward's geological casts. The sum required for the casts — sixteen hundred dollars — was obtained almost entirely from former members of the seminary, by the persevering efforts of Miss Edwards; and the large expense of setting up was generously met by Mr. Williston. The art gallery, consisting of a large central apartment with side alcoves, and two smaller rooms, occupies the entire upper floor.

The joyful gathering at the dedication of this long-desired building proved to be the last time Deacon Porter ever came to the seminary. Though he then seemed as well as usual, and in full sympathy with the gladness of the occasion, it became evident during the winter that a disease of the heart, from which he had long suffered, was advancing to a fatal termination. Writing to Miss Ward early in February, he alluded to his state of health, and added, "I have entire confidence that my great Physician understands my case, and will bring me out all right." But it was not recovery for which he was looking. Difficulty of breathing seldom allowed him to lie down, by night or by day; and nearly a week before his death, paralysis of the left side rendered him speechless and unconscious. But a day or two later, as he lay in his reclining chair with closed eyes, apparently beyond intercourse with earth, those about him observed that he was moving his finger as if writing. Quickly comprehending the sign, they placed a pencil in his hand, and paper beneath it. Yes, he was trying to write; the faltering hand, which the eye could no longer guide, traced the words, "Messages of love to all seminary daughters." Repeating the effort, as if doubtful whether he had succeeded, he wrote, "Love to all pupils of seminary." One of the eldest of the "daughters," Mrs. Foster, was by his side, and read aloud what he had written; he perceived that he was understood, and made a sign that he had done. That penciled message is treasured among the most sacred mementos which the seminary possesses of its departed friends. "He was still alive when it reached us," says the journal; "perhaps you can imagine how touching was the sight of those last tender words. Oh, if we could only have sent back our grateful response, — if some loving message from the seminary daughters who owe him so much might have overtaken the departing soul! But he was beyond our farewells." So those last days went by, save that once again he wrote a few words, brokenly conveying to Mrs. Porter that he was perfectly at peace. Two days after completing his eighty-second year, on Sabbath morning, March 4, 1877, he fell asleep.

"Next to Mrs. Porter," wrote the principal, "Mount Holyoke Seminary is chief mourner." For more than forty years this faithful friend had loved it with a father's love; his one anxiety, in the serene evening of his days, had been to see others come forward to undertake with like fidelity, the self-sacrificing work that he must lay down. His desire had been granted; in a letter two years before his death, he had spoken of the comfort it gave him to have "two such men as Mr. Williston and Mr. Sawyer, so competent and so willing to afford us their aid, at a time when they are so much needed. . . . It is about time for friends of the seminary to give up their distrust of God's faithfulness. I have been thankful every time I have had the seminary on my mind lately that we have these new helpers to lean upon."

Much might be said of the personal history and character of this revered friend, and of the providential training that made him what he was. But the story of his upright and manly youth, his energy and success in business, his early and systematic charities, and his Christian activity in the communities where he resided, cannot here be told. He was a steadfast friend and helper to Monson Academy, and Amherst College; and was also for many years a corporate member of the American Board of Commissioners for Foreign Missions.

Dr. Tyler, who had long known Deacon Porter as a fellow trustee of the seminary, wrote: "My association with him has left upon my mind the liveliest and deepest impression of the wisdom of his plans and counsels; the simplicity, purity, and utter unselfishness of his character; and the unspeakable value of such a practical, sensible, prudent, efficient Christian man of business in the founding, rearing, and managing such an institution. As genial as he was generous and just, as remarkable for his courtesy and pleasantry as he was for his evangelical faith and puritanical piety, he has ever held the unbounded confidence and affection of trustees, teachers, and pupils; and I have rarely seen so beautiful a sight as this childless old gentleman gladly welcomed by his children and grandchildren, as they gathered about him in loving and joyful groups, whenever he visited the seminary."

His love for the young, and sympathy with them, was indeed a winning trait. Children might well have been a little awed by his commanding figure, but for the kindly and benignant face and the playful twinkle of the eye, which made it plain that his heart was with them. In 1872, he had married Mrs. Mary Stafford, a cousin of the first Mrs. Porter; and during his remaining years, the seminary daughters who came and went always found there so delightful an atmosphere of loving companionship and tender ministries, that there seemed nothing more to desire for him on earth.

In the catalogue for 1876-7, for the first time, modern languages found a place in the prescribed course, one term of either French or German being required. A full half-year is now devoted to the language chosen; and a beginning thus made, students are encouraged to pursue it much farther, if circumstances permit. At the date above named, the history of art also was assigned a place in the senior year; and Miss Blanchard, then in Europe, prolonged her stay, in order to give as much attention as possible to whatever might be helpful in teaching it. In the spring of '78, Prof. William H. Goodyear — then connected with Cooper Institute, and now curator of the Metropolitan Art Museum — gave a finely illustrated course of lectures on the history and philosophy of art, which was highly appreciated and enjoyed. There is ordinarily a course in this department each year, given either by Professor Goodyear, or Professor Mather, of Amherst College; the admirable lectures of the latter are devoted mainly to sculpture.

It was this year that the custom arose of having occasional concerts of a high order given at the seminary by professional musicians. Ever since, there have been two or three of these every year, in addition to the musical entertainments given by the teachers and pupils of the department. They are much valued as a means of culture to all, and particularly to those who are making music a special study.

In 1878 there was published a catalogue of the Memorandum Society, for the forty years ending in 1877. It was prepared by Mrs. Mary W. (Chapin) Pease, whose thorough acquaintance with the history of the seminary, and with its successive classes for the first thirty years, gave her peculiar fitness for the work. The society was organized by Miss Lyon in the first year of the school, for the purpose of obtaining and perpetuating facts in regard to the history of its members. It included not only graduates, but other pupils who wished to join. The payment of a small initiation fee secured membership for life, and entitled to the society's publications. Once in five years a catalogue was issued, stating the chief facts in the history of each member since the preceding report. At this time, the names of all who had ever joined the society were included, with the history of each from the time of her becoming a member to date; or, if deceased, till her death. The whole number was 2341, of whom 1604 were graduates. The labor involved was immense. To trace the history of those who had failed to report, and who had in most cases changed not only their residences but also their names, was no trifling task. But there were only nine members of whom no information was obtained; and most of these were heard from soon after the catalogue was published. Of the 2341 members, over 1900 were still living.

At the Paris Exposition, the seminary contributed to the educational exhibit of the United States, by special request of Dr. Philbrick, the commissioner. The time for preparation was so limited that the seminary was not adequately represented, except perhaps by its printed documents. A second large supply of catalogues, and of the historical sketch, was soon sent for by the commissioner, who afterwards said to the correspondent of the *New York Tribune*, "Perhaps no exhibit in the educational department excited more attention than that of the higher education of women, represented by Vassar, Wellesley, Mount Holyoke, and the Georgia Female College." A medal was awarded to the seminary, which may be seen in one of the cases in the south alcove of the art gallery.

The death of Hon. Edmund H. Sawyer, of Easthampton, in November, 1879, removed a warmly interested friend of the seminary, still in the prime of life. For six years he had been a trustee, and generally a member of the executive committee, where his ability and experience were highly prized. Though burdened with many public and private responsibilities in addition to his own business affairs, he was ready to give time and labor to seminary matters whenever needed; the previous summer, though far from well, he had attended to the duties of the treasurer, then in Europe, and had superintended the various improvements made in the seminary buildings and grounds. Both in public and private life he had filled many responsible positions; and in them all had been honored and loved as a noble Christian man. The following sentences are quoted from the resolutions passed by the trustees:—

In his death, the seminary loses a warm friend, a generous benefactor, a wise counselor, a faithful guardian, especially of its financial interests. We cannot forget that in him the state also has lost an incorruptible and influential legislator, as well as a good citizen; the church an exemplary and active member; and the family an affectionate husband and father.

Among the improvements made in the years 1879 and 1880, two were of special importance. The first of these was the artesian well. In dry seasons, there had sometimes been a scarcity of water for laundry and other work, the supply for these purposes coming mainly from the brook in the grounds. As there seemed to be a prospect of easily relieving the difficulty in this way, boring was begun near the southeast corner of the gymnasium. The undertaking was not so soon accomplished as had been hoped, and once came near being abandoned; but at the depth of four hundred and fifty feet, water came at the rate of forty gallons a minute, and was found to be of excellent quality. As it rose spontaneously only to within sixty feet of the surface, a small steam engine was provided for pumping, which has since been made to do service also in various other ways. From that time, the supply of water has been abundant and unfailing.

The next enterprise was the elevator. Years before, it had been spoken of, but only to be dismissed as out of the question. There is mention of the subject on the minutes of the trustees for 1872, a report having been made that "the matter of an elevator cannot now be considered as practicable." But the need continued to be felt; and as years went on, new contrivances prepared the way. It was early in 1880 that the project was really taken in hand. By that time, it was well understood that a natural water power was not essential in order to have a hydraulic elevator; and since it was clear that the thing desired could be done, it only remained to raise the money and do it. The first contribution taken up for the elevator is thus mentioned by one who shared in it: "It was when I sat at Miss Shattuck's table, in the spring of 1880. After talking ourselves into thorough earnestness, we agreed to come to the next meal bringing each five cents. Sixty-five cents only was the result, for two of us forgot that all important beginning." Yet the little collection answered its purpose. As in the days when steam heating was aspired to, so now, there was much letter writing to home friends and to former students, and many visits were paid to those within reach who were known to be liberal. Miss Shattuck had long been deeply interested, and was active in soliciting donations, as also were Miss Ward, and other teachers. So much success attended these efforts that the enterprise was not long a doubtful one; at the meeting of the trustees in June, the executive committee were authorized to proceed, and early in the autumn the elevator was running. Its cost was $4,273.50, all of which was given for the purpose. It is of the best make, and the handsome car accommodates about twenty passengers. The benefits expected from the elevator have been fully realized, particularly in making the rooms in the upper stories as eligible as any in the building; it has been run without accident,—accident being indeed hardly possible,—and with little expense.

In this same summer of 1880, it became known that the observatory, so long desired, was about to be built. At first, a little mystery surrounded the matter; but means having been promised, no time was lost in beginning. The seminary was especially favored in having

THE INTERIOR OF OBSERVATORY.

Equatorial Telescope and Spectroscope.

on its board of trustees the eminent astronomer, Dr. Charles A. Young, of Princeton, who had for many years been its lecturer. To his judgment everything relating to the building and equipment of the observatory was referred; and he generously bestowed upon it time and thought as if it had been his own. No suitable site could be found within the seminary grounds, the view of the heavens being everywhere more or less obstructed by trees or buildings. Across the street, however, an excellent location was found, southwest of the high school building, where an acre was purchased by the friend who had undertaken the matter, and whose name by this time was known.

In the Smithsonian Report for 1880, in an article on the astronomical observatories of this and other countries, the one at this institution is thus described by Dr. Young:—

Through the liberality of Mr. A. L. Williston, of Northampton, the seminary has recently been enabled to erect a small but very complete astronomical observatory, supplied with all the necessary instruments. It is designed to furnish the means for instruction to any who may wish to make the subject a specialty, and to give opportunity to any of the teachers or post-graduates who may take an interest in astronomy to make observations of real value.

The building consists of a tower with a dome eighteen feet in diameter, flanked by two wings, one extending to the west and one to the north. The dome is very light, and rotates so easily that any young lady can manage it without difficulty. The arrangements for opening and closing the shutters which cover the slit in the dome, and the openings for the transit and prime vertical instrument, are worked with equal facility. In the dome is mounted a fine eight-inch equatorial by Clark, completely fitted out with clock-work, finding-clock, micrometers, spectroscope, solar eye-piece, etc., and so arranged that the circles can be read and the clamps and tangent screws worked from the eye-piece of the instrument. The object-glass is almost entirely the work of the senior Alvan Clark, and is one of the most perfect specimens of his art.

In the transit-room is mounted a meridian circle by Fauth & Co., of Washington. The instrument has a telescope of three inches aperture, and circles of sixteen inches diameter, reading to seconds by two microscopes. It has a reversing apparatus, and is fitted with a "latitude level" and micrometer, so that it can, if desired, be used as a zenith telescope. A large collimator is mounted upon a pier south of it, and in the corner of the room is a clock with Denison escapement, also by Fauth & Co., as is the chronograph, which is mounted in an adjoining closet. The observatory possesses also a sextant and artificial horizon, and a set of meteorological apparatus.

There is no instrument in the prime vertical room,—which is used in connection with the study, for recitations,—but it is provided with a pier and shutter, so that the meridian circle can be set up there if it is ever thought desirable to make observations in that plane.

The cost of the whole was ten thousand dollars. The chief part of the expense was for the instruments, as a massive structure was considered undesirable.

When the next anniversary arrived, all was complete. At the close of the exercises in the church, the returning procession moved on past the seminary to the new observatory, to attend the brief ceremonies of its dedication. Around its entrance gathered a joyful throng, —trustees, teachers, students, and guests. Dr. Young spoke of the history of the enterprise, and the objects in view. His excellency, Governor Long, followed in a neat and appropriate speech, and the dedicatory prayer was offered by Dr. Seelye, of Amherst.

A peculiarly tender and sacred recollection is inseparably associated with this one of the seminary buildings. It was the gift of bereaved parents, in memory of their eldest son, John Payson Williston, a manly and beautiful boy, who died April 23, 1879, at the age of fourteen.

The observatory was ready for use in good time for a great astronomical event—the transit of Venus in 1882— on which occasion valuable observations were made. The longitude of the observatory was also determined. The solar eclipse of March 16, 1885, was successfully observed, and the report of Miss Bardwell was communicated by Dr. Young to the *Sidereal Messenger*. Meteorological observations are regularly kept. In connection with the required study of astronomy, the telescope is freely used for exhibiting to the pupils sunspots and solar prominences, the moon, planets, double stars, and nebulæ. Those who take advanced work have instruction and practice in the use of the sextant, meridian circle, equatorial telescope, and spectroscope.

A destiny almost romantic awaited the old telescope. It was still a good instrument of its size, though the apparatus for managing it was somewhat inferior. Miss Ferguson, of the Huguenot Seminary, desired to obtain it for her school, as the wonderfully clear atmosphere at the Cape of Good Hope is so favorable for astronomical study. Accordingly it was purchased from the seminary by Mr. Williston, provided with the needful accompaniments, and forwarded. The American astronomical party, sent thither to observe the transit of Venus, selected their station at Wellington, in the seminary grounds. During the weeks of preparatory drill, Professor Newcomb trained some of the teachers along with his own assistants; and when the transit took place, the old Holyoke telescope, with a Holyoke graduate behind it, added a valuable observation to those made by the various members of the party.

The grounds of the institution were greatly enlarged by additions in 1880 and 1881. For some years, Miss Ward had considered it desirable that the seminary should secure a large tract including the summit of Prospect Hill; the charming views it commanded were of themselves a sufficient attraction, and its rare possibilities for landscape gardening could not be overlooked. Here, the seminary might in time have groves of its own, safe from the woodman's axe. A six-acre lot on the hillside, containing a few fine old trees, had been purchased at her suggestion for the seminary by Hon. E. A. Goodnow, of Worcester, some time before. In 1880, the chief part of the hill, comprising from twenty to twenty-five acres adjacent to that already owned, was presented to the institution by the same friend. To the gift of the land he added two thousand dollars to be used—together with smaller sums from other friends—in improvements; and later, provided a permanent fund of five thousand dollars whose income should be devoted to the same purpose. This portion of the seminary estate bears the name of Goodnow

Park. Thousands of young trees have been planted, and already its growing beauty attracts many to try the pretty winding walk up the hillside, or the drive leading from the iron bridge by a more gentle ascent through the southern part of the park, to the pavilion that crowns the summit.

In July, 1880, Colonel Rice, of Conway, passed from the labors of earth to the rest of heaven. He had lived to a good old age, and at the time of his decease lacked only two days of completing his eighty-sixth year. In May previous, replying to Miss Ward's inquiries about his failing health, he had expressed his deep interest in the seminary, and his confidence that it would continue to increase in usefulness and power. He was remarkable for the symmetry and consistency of his Christian character. While firm in his adherence to principle, he was most kindly and genial; in business matters he was active, enterprising, and judicious, yet a liberal and systematic giver, who loved to help the Lord's work or minister to the poor. Humility was joined with an unfaltering trust; and the Friend with whom he had walked through life did not fail him in death.

The death of Mr. Kingman, another valued trustee, occurred somewhat suddenly November 1, 1880, at the age of sixty-six. He was a native of Providence, but had long resided in Boston. He was an earnest and devout Christian, actively interested in many kinds of work for the Master. He had been a member of the prudential committee of the American Board of Foreign Missions for many years, and a trustee of the seminary from 1856 till his death.

At the next meeting of the trustees, they put on record their tribute to the memory of these lately of their number, from which the following passage is quoted:—

> Deacon Kingman had been a member of this board twenty-four years, and Colonel Rice twenty-two; a longer time than any of the trustees who now survive, and longer than most of those who preceded

them. During all these years, they have been constant almost without interruption, in their attendance at the meetings of the board, and the seminary has found in each of them a faithful servant, a wise counselor, and devoted friend. Much of the time Deacon Kingman has been also a highly valued member of the executive committee and committee of finance. Both of them have been unsparing of time, money, and prayerful, painstaking service.

The death of Mr. Durant, a trustee of the seminary from 1867 to 1878, took place October 3, 1881. Though his absorbing labors in behalf of Wellesley College had of late prevented him from visiting the seminary, it still held a place in his heart and his prayers. His labors for the religious welfare of its students are held in grateful remembrance; and the library which owes so much to him as well as to Mrs. Durant will ever be their memorial here. The trustees passed a resolution, at their next meeting; from which a few sentences are quoted:—

In the recent decease of Henry F. Durant, Esq., we mourn the death not only of a munificent patron of woman's higher education, but of a former trustee, wise counselor, and liberal benefactor of Mount Holyoke Seminary. . . . We sympathize deeply with the trustees, teachers, and students of Wellesley College in the loss of its founder, and tenderly with Mrs. Durant in her sore bereavement.

The religious history of this period was not unlike that of the preceding one. Each year, a large majority of the pupils were church members, and many others classed themselves as Christians, though they had not made a public profession of their faith. Of the rest,— usually not more than one-fifth or one-sixth of the whole school,—in most cases the greater part were hopefully converted before the close of the year. The work needed in behalf of the large number of young disciples, who, however sincere, are inexperienced and weak, is always great. Its progress is generally noiseless and unobserved, yet real, and at length evident. "So is the kingdom of God as if a man should cast seed into a field, and should sleep and rise night and day, and the seed should spring and grow up, he knoweth not how, . . . first the blade, then the ear, after that the full corn in the ear."

A few extracts from the journal, and from other records made at the time, will illustrate what has been said. One point specially noted in regard to the year 1872-3, during which there were twenty-five or thirty cases of hopeful conversion, was "More than usual growth among Christians." While Dr. Kirk was at the seminary in the spring of that year, many asked to converse with him; the journal says: "There seemed to be earnest desires to learn more, and go on faster in the ways of the Lord. Dr. Kirk, whose labors had in the past been chiefly for the impenitent, said he felt more than ever before the importance of the work to be done for those already Christians, but weak."

In the spring of 1877, on a special day of prayer, it was said: "All the meetings were very encouraging, especially one to which were invited only those who desired to attain fuller personal consecration to Christ. The lecture room was crowded; a great many short voluntary prayers were offered, which seemed heartfelt. The religious state of the family has on the whole been encouraging all this term. . . . There are now in school only eleven who do not call themselves disciples."

In 1878,—"Soon after the week of prayer, the young ladies began to have many little meetings by themselves. Then they asked that they might hold a general meeting twice a week in the lecture room. It occupied the fifteen minutes before tea, on Wednesdays and Sundays. These hopeful tokens continue, and the meetings are full and interesting."

In February, 1881,—"The influence of the day of prayer for colleges has been manifest; Christians have been quickened, and we rejoice over some who have come to Jesus for the first time. But we need—oh, how greatly—that which is so much needed by the churches everywhere,—more earnestness, constancy, and consistency in the lives of professing Christians."

In June, 1882,—"There were, toward the close of the year, some conversions that awakened especial gratitude, because these were souls long prayed for."

Several of the teachers during this period left the seminary for service in foreign lands. The first was Miss Annie M. Wells,—a graduate of the class of '67, and for four years a teacher,—who in 1874 went to South Africa, to assist in the lately established Huguenot Seminary. In 1876, Olive J. Emerson, M. D., a graduate of '65, who took her degree in medicine at Michigan University in '74, went to Tavoy, Burmah, as the wife of Rev. Horatio Morrow, missionary of the Baptist Union. She had been highly valued as the seminary physician for two years; and in her mission field she ministers to the needs of the body as well as the soul. Her example was followed by her successor at the seminary, Adaline D. H. Kelsey, M. D., who graduated at Mount Holyoke in '68, and at the Woman's Medical College of the New York Infirmary in '76. She went in 1878 to Tungchow, China, as a missionary physician of the Presbyterian Board. In 1877, Miss Susan M. Clary of the class of '63—for fourteen years a teacher at the seminary—went to South Africa, to begin a school for girls at Pretoria in the Transvaal. Within a year her work was done. A severe attack of pneumonia was followed by consumption, of which she died August 3, 1878.

Miss Clary seemed remarkably well fitted for the enterprise which she had so successfully begun. She had devoted herself to it for life,—the long life on which she thought she might reasonably count. To a friend who inquired how soon she meant to come home for a visit, she replied, "Not for forty years." Always well and cheery, she enjoyed having much to do and to oversee. Naturally quick, resolute, and systematic, she accomplished a great deal, and had time to spare for any unforeseen demand. She was an enthusiastic teacher, and delighted also in loving ministries to those under her care. One of the old scholars says: "I often thought Lowell might have taken her for a model when he wrote

> 'She doeth little kindnesses
> Which most leave undone or despise,
> For naught that sets the heart at ease,
> Or giveth happiness or peace,
> Is low-esteemed in her eyes.'

. . . I think her part in heaven must be that of a ministering spirit; she could not be happy unless she were doing for others."

In June, 1882, Miss Ward offered her resignation, the state of her health for years having been such as to justify and demand release from the heavy responsibilities of her position. Though her love for the work was not less, her strength was no longer equal to it. Unwilling that her connection with the seminary should be terminated without previous trial of a long absence, the board prevailed upon her to adopt this course. During the winter, she journeyed as far as California, where she remained many months, but without marked improvement. At the next anniversary, the trustees accepted the resignation again presented, and passed resolutions from which the following paragraphs are quoted:—

Voted, That able, faithful, and laborious service for eleven years, in which, notwithstanding unprecedented competition and other adverse circumstances, the seminary under her administration has not only continued to prosper, but increased in numbers, enlarged and beautified its grounds, multiplied its buildings and educational advantages, and raised the standard of attainment in literature and science without letting down at all the moral and religious standard, entitles Miss Ward to the grateful recognition of her services by the trustees and by all the friends of Mount Holyoke Seminary.

Voted, That we sympathize with her in the protracted ill health which has necessitated her resignation; and assure her of our best wishes and earnest prayers for her speedy restoration to perfect health, and that she may have many years of still more useful labor in the cause of woman's education.

On anniversary day, Dr. Tyler read at the collation a paper reviewing the growth of the seminary during the administration just closed, from which a few sentences are quoted:—

All these improvements and signs of progress have been introduced within a single decade. . . . The whole sum which the seminary has received and expended in such improvements as these during the principalship of Miss Ward is over eighty-six thousand dollars. Meanwhile there has been real progress in the course of study and the methods of instruction, particularly in those branches of science and art which are represented by the Lyman Williston Hall and the astronomical observatory; and in the establishment of a post-graduate course in ancient and modern languages. . . . Mount Holyoke Seminary means to prove all things, to hold on to all that is good in her past, and to reach out after more and better things in time to come. Her motto is substantially that which President Stearns was so fond of repeating for Amherst: The best education in literature, science, and art; and all for Christ.

While this period had been marked by great progress in material things, they had been sought mainly because the highest usefulness of the seminary to the cause of Christ was involved. Had the need of the hour failed to be met, the institution would gradually have ceased to attract the class of students most able and disposed to do good in the world. "I would a thousand times rather see it blotted from the face of the earth," said Miss Ward, "than to see it secularized. It is my deep conviction that the highest intellectual and religious culture must here exist together. May it never be that the calls for Christian workers, more frequent and pressing now than at any time in the past, shall fail to find a response here."

The journal of June, 1883, after alluding to the resignation of Miss Ward, added: "But she is ours still, and ever will be. . . . We owe her very much; we do not forget it, as we earnestly ask for her many years of health and of abundant usefulness, spent in the clear light of God's presence. 'Give her of the fruit of her hands, and let her own works praise her in the gates.'"

CHAPTER XVII.

ADMINISTRATION OF MISS BLANCHARD.

FROM 1883.

DURING the school year 1882-3, there was still hope that Miss Ward might be able to return, and the school was meanwhile in the care of the associate principals, Miss Blanchard and Miss Edwards. The attendance was large, and the year was pleasant and prosperous. Several important acquisitions and improvements marked its progress. Among these was the purchase of the estate next the seminary grounds on the north, once belonging to the Dwight family, and afterwards to the late George Chamberlain, Esq. Besides a large dwelling house, it included seven acres of land. The house, after some alterations and repairs, was principally devoted to the classes in drawing, and to music rooms. The need of the latter had been felt for some time, as the number of those taking private lessons had increased. And the grounds became the more attractive by so large an addition.

In the autumn of 1882, a small greenhouse, costing about five hundred dollars, was built at the south end of the gymnasium. It has not only been a constant source of enjoyment, but has also furnished a winter home for many delicate and beautiful plants from the botanical garden. It was provided by the thoughtful kindness of a former member of the seminary, Miss Emma E. Dickinson, of Fairport, New York. Another, Mrs. Ruth (Washburn) Bancroft, of Buffalo, a pupil in Miss Lyon's time, sent five hundred dollars for renovating the seminary hall; and this having been done, the young ladies themselves secured funds for replacing the old settees with chairs. A surplus remaining from their gift was appropriated to improvements in the reading room. There were important donations to the art gallery, including paintings, casts of noted sculptures, and Chinese bronzes of great antiquity and value.

There were during the year several special courses of lectures, among which was one by Dr. William T. Harris, on the philosophy of education, a short historical course by Dr. John Lord, and one on biology by Professor Rice, of the Wesleyan University, at Middletown, Connecticut.

At their annual meeting, June 20, 1883, the trustees passed a vote of thanks " to the associate principals, Miss Blanchard and Miss Edwards, who have administered the government and discipline so wisely and successfully during the absence of the principal." The resignation of Miss Ward having been accepted, Miss Blanchard was unanimously elected principal, with Miss Edwards as her associate, which offices, respectively, they continue to hold.

The period has been one of prosperity and growth. The attendance has been large; and though the building of a house for the steward has left free for students' use several additional rooms in the south wing, it has not been practicable to receive nearly all the well qualified applicants. This year, 1886–7, all the available rooms in the Dwight house have been used for dormitories.

Early in 1884, the front parlors were handsomely refurnished by the generosity of some of the trustees, who kindly took upon themselves not only the expense, but also the superintendence of the matter. To these parlors the young ladies are accustomed to resort at pleasure, during the recreation hours before and after tea.

The paper mill property was purchased in the summer of 1884. This secured the control of its water power, should it sometime be needed for making electric light, or for any other purpose; and also of the supply for watering the lawns and botanical garden. The pavilion in Goodnow Park was built the same year, and the house for the steward the summer following.

In 1886-7, the library building was enlarged by adding at the north end another room, nearly as large as the first. It is occupied by cases compactly arranged, which will shelve twenty or twenty-five thousand volumes on one floor; and is capable of being so modified as to hold many thousands more, should it by and by be needful. Double fire-proof doors separate it from the original room, which with its eleven or twelve thousand volumes continues to be a constant resort during thirteen or fourteen hours of the day. There is a classified index to the library, as well as a card catalogue for public use, besides the official card catalogue. As yet, the library has only the nucleus of a permanent fund, but there are annual appropriations from the general fund for the purchase of books.

Besides numerous gifts for the art gallery and cabinets, there have been important additions to the apparatus, particularly to that for microscopical study. One friend gave a thousand dollars for this purpose, in 1885; and over three hundred had been sent a short time before by alumnæ in Boston and Worcester. There are at present twenty compound and twenty-four dissecting microscopes, with a fine microtome and other accessories, in daily use by enthusiastic students and teachers.

In regard to the course of study, the requirements for admission have been somewhat increased, and modifications made by which time is secured for further progress in certain departments. Text books and works of reference are supplemented by frequent lectures from teachers, in addition to the regular courses given by the non-resident lecturers. In 1883, Professor Mears, of Williams College, became the lecturer in chemistry; and in 1885 Professor Kimball, of the Worcester Technical Institute, the lecturer in physics. Besides the regular lectures, there have been short courses in biology by Professor Wilson, of Bryn Mawr College, and Professor Sedgwick, of the Massachusetts Institute of Technology; also in political science by Edward W. Bemis, Ph. D.

For years, the teachers of various departments have from time to time availed themselves of the advantages of the leading technological institutes, and have also had private instruction from specialists in their respective studies. Physics, chemistry, botany, and zoölogy have been thus pursued in repeated instances, not to mention the special study of mathematics, psychology, and ancient languages.

The amount of work done by students beyond what is required has for several years been constantly increasing. In 1884-5, eighty did advanced work in languages, literature, mathematics, or sciences, during a part or the whole of the year; in 1885-6, the number was over a hundred. In most cases, the time thus devoted—perhaps a year or more—is taken before completing the regular course, rather than after graduation. Laboratory work, in one line or another, is found to have strong attractions for many; in some cases, it has been pursued for two or three years beyond the regular curriculum. The number of hours per week thus spent has varied from three or four to twelve or fifteen. In each department, work in advance of the course is encouraged, and provided for as far as desired.

The following extract from one of the journal letters gives a good picture of seminary life at the present day. It was written in June, 1885, by Miss Bowers:—

If you were here to-day, this rare June Wednesday, your first thoughts might be for the beauty out of doors rather than for friends within. Can we help you to see it? Walk with us up the winding footpath to our pretty pavilion in Goodnow Park on Prospect Hill. How beautiful the noble avenues of trees that used to bound our grounds; the little brook and quiet pond where the boats "float double," boat "and shadow"; the old chestnuts around us, and the infant trees for our future groves; the picturesque old cider mill in the field to the east, and the grist mill down by the pond; the encircling hills, and the gates ajar between Holyoke and Tom to let the sunset glory through. Descend by the carriage road that winds up from the brook toward the south and around the pavilion at the summit of the hill. The meadow is full of buttercups and daisies as of old, and the little foot bridge by the wheel house is the same, but the boat house below is a new resident, and in the basin above the bridge there will soon be a fair white host of water lilies. Pause awhile, if you will, among the sweets of the botanical garden, then cross the grass to Miss Lyon's grave.

The English ivy, which covers the ground of the enclosure, is somewhat winter-killed each year, but makes afresh a rich green carpet each summer. How beautiful this little grove, and how merry the occupants of the hammocks and the grass beneath its shade! Answering voices float up to us from the boats on the water, and across the grass from the lawn-tennis players, under the drooping branches of the great black walnut tree, the royal resident on our domains. A little nearer us, victors and vanquished are laughing over a game of croquet. Girls are scattered about the grounds, singly or in groups, busy with book or needle work or pleasant chat, and we know they are taking in fresh life for nerve and heart and brain. Sweet recreation days in field and wood and on these beautiful grounds! They must be, beyond our power to estimate, "faire gospellers" of peace. Even the walk from the house over to Williston Hall must compose one's mind for a recitation there! The supper bell calls; the evening brings all home to quiet and sleep, from which we wake into a study-day world. Will you stay through its hours?

There is no bell in the morning until the rising bell, which is itself somewhat tardy in rising, as compared with the old standard, and has a subdued tone of voice, as if afraid it may wake somebody. Girls must be prompt at breakfast and quite ready for it, but they may choose their own time of rising, if it be not too early!

Look about in the basement a little, after breakfast. The elevator, you remember, cuts out one corner of the middle room, and the bread closet occupies another. The bread cutter close at hand; the great oven in the domestic hall; the closet built against the oven, with rows of shelves, whereon all our bread is placed to rise and kept moist by a tiny spray of steam; the big broilers for steak; the former oven converted into a great plate warmer; the steam kettles which supply so easily the needful quantity of coffee or chocolate or soup; the soapstone sinks for dish washing; the flour sifter; the refrigerators, for butter, for milk, for other food,—all these make work easier or furnish our table more healthfully, and something is brought in every year to these ends.

The bell for morning devotions calls us to the seminary hall, made very attractive now by the furnishings of a few years past. Miss Lyon, Miss Fiske, Miss Hopkins, Mrs. Eddy, Deacon and Mrs. Porter, Mr. and Mrs. Samuel Williston, look down upon us from the walls, and choice photographs are there besides. Miss Edwards is our preacher, and we doubt if many pulpits are filled more acceptably. Can girls go away without receiving life-long impressions for good from that desk? We have responsive Scripture readings in these morning exercises, and

THE BOTANICAL ROOM.
Lyman Williston Hall.

it is surely nowhere better done; the many voices are as one. The half-hour over, let us go first into a botany class in Williston Hall. A lesson has been assigned from the text-book, but a part of the hour will be occupied by a lecture on the same topic, giving what the book does not. Notes must be taken and reported the next day, when they have also a lesson on another assigned subject. One side of the recitation room is taken up by the cases and drawers of the great herbarium; botanical specimens of various kinds, that are hung all over the other walls, make the room as bright as a gallery of paintings. Little tables are here whereon to use the dissecting microscopes and to make the drawings required. The great work room of these classes is the world in the open air, and especially the botanical garden, where you may see them at almost any hour of the day, intent on finding out all the family secrets of the garden dwellers. You know, I think, that a part of their required study comes in the autumn, for the sake of "the golden rod on the hill, and the aster in the wood, and the yellow sunflower by the brook," and others of their kin. The recitation over, the bell-glasses are removed from the compound microscopes, by the windows, and the "spirogyra" girls are at them. Miss Hooker calls them so because the students beginning botany in German universities— who do the same work—get the appellation of "spirogyra men." All winter, these girls were studying dried specimens, or mosses, ferns, and lower plants from Wardian cases. The life history of the fern, tracing its development from the spore, making drawings of all stages of growth, cutting sections of stems to study tissues, analyzing all the North American ferns they can obtain, following out much the same plan in the study of mosses and grasses—this has been the work of most of them. One Saturday we were summoned excitedly by a radiant student to see something which the microscope had caught just at the right moment—moss antherozoids in motion—a rare sight. Hasty feet sped over to Williston Hall as long as the show lasted; but an irregular procession of girls was moving in an opposite direction at the same time, for the sun was greatly disturbed, and they went to the observatory to witness the "prominences" which were unusual and wonderful in variety and form. The infinitesimal and the vast, too small, too large for our limited eyes to see, were brought curiously near together before our awed and admiring gaze that day. The observatory work is in the hands of Miss Bardwell, an enthusiastic student as well as teacher, who does all she can to make the "spacious firmament" itself an illustrated text-book for her classes. She is our herald, to give tidings from her watch tower of new-comers in the fields of space, or of changes there among the shining host already known. Whenever there is anything of unusual interest to be shown, she spends many hours at her post, aside from the time given to her classes. One might well wish to be a pupil again, in one of those classes, for the sake of the nightly visions of "other worlds than ours."

Now let us turn from "the heavens above" to "the earth beneath," with the zoölogy class. They're intent on earth-worms to-day, "a good typical form"; the general anatomy, the structure, the development, as revealed by the microscope, are studied thoroughly. Another day we should find them at work on clams or on lobsters in the same way; upstairs, in the bird alcove, we shall find others analyzing unmounted birds, kept for such use. Come to the work room in the basement and see what some of the advanced students are about. It must be the medically inclined who enjoy so much these dissections. They, as well as the students in the regular classes, are required to make many drawings. One has mounted the skeleton of a cat, going through every step of the process herself. Miss Clapp's "retainers" and offerings are many. The ferryman sends tidings of a cuckoo's nest by the ferry; one boy brings very large earth-worms, another some field-mice, or a snake, or a bird's nest; three infant rabbits, ploughed up in a field, have afforded us all much amusement. Observations in the quadrangle lead to the belief that every girl in school has been laid under bonds to bring a mud turtle there, whose end is biology. And a cat sometimes comes—alas, poor pussy! The zoölogy room is used by the physiology classes also—did you notice the models, really beautiful in their way, of eye, and ear, and throat? One has admirable opportunity to study comparative anatomy from the skeleton and manikin here, and the dissections in that basement room. There are no recitations in physics and chemistry this term,* so we will leave the laboratories and apparatus, with their manifold facilities, and go for a few moments to the rooms for geology and mineralogy. One wants a long time to see the treasures in the five rooms, but perhaps the greatest of them in the eyes of Miss Cowles, at present, is the lithological microscope in her class room, so lately given by the Worcester Alumnæ Association. She had an enthusiastic class in mineralogy through the winter, who took it as an optional, and they enjoyed greatly the crystallography which the microscope helped them to know. Looking into that tube sometimes was like reading the twenty-first chapter of the Revelation, and one forgot science in such visions of splendor and purity.

We want to take you over to our two-year-old house for drawing and music—the old Dwight place. You may stop to admire the smooth-shaven lawn between it and the seminary, with its scattered trees and shrubs. The pupils do a great deal of charcoal work; here are some lovely little landscapes, taken from nature; many are copying from casts; some are making portraits from photographs or from life. Coming back to the great house, let us step into the elevator and go up to the rooms for painting, at the north end of the upper floor. The work is done in both oil and water-colors, and is mostly fruit and flowers.

It is time for us to descend from high art to our dinner in the basement, yet we think you will agree with us that something of high art

---

*Much advanced work is done in chemistry, chiefly in qualitative analysis. The small physical laboratory lately fitted up affords increased facilities for experimental work, in which an extended course is arranged. More room is greatly needed in this as well as the other departments of natural science.

has been here also, in the making of this delicious bread and biscuit, if in nothing else. The girls made it all, and they have prepared all this dinner under the supervision of the genial and considerate matron, who finds them almost always "most helpful and pleasant to work with." And they will have a good time over the dish washing by and by; if you don't believe it, stay one day after dinner and use your eyes and ears. But they will not do all the work that their mothers did when here, neither will they on Wednesday. It has seemed best to make changes from time to time, giving some of the harder parts of the domestic work into the hands of women who come in from the village; the man who took the place of Cornelius gives all his time to the baker's and other kitchen work; the "moping circle" that a girl once wrote home about, is a thing of the past, unless the use of a dish mop entitles one to a place in it. There is still abundant opportunity to learn promptness, efficiency and care-taking for the common good. Newspaper and magazine articles, and cooking schools for the wealthier classes, are at one with seminary teachings more than ever, now, in dignifying housework by bringing into prominence its scientific and æsthetic aspects; we want these girls to enter also into the higher spirit that says with George Herbert,

"Who sweeps a room, as for Thy laws,
Makes that and the action fine."

It is time for the afternoon recitation hours—only two now, and only three in the morning, because more time is given to each. These history classes have gathered their store of knowledge from their richly furnished alcove in the library, using printed topic books prepared by their teachers; but you will find that they have been to the reading room as well, and know something of the Greece and Rome and England of to-day, as well as of the centuries gone.

If the English literature class is discussing Lady Macbeth's or Portia's character, we think we can promise you some bright and discerning remarks of their own, not the commentators'; if they give you some of Bacon's Essays in their own or in his words, you will be glad that they have such weighty antidotes to the light reading which most girls take in such large and frequent draughts. It is an encouraging symptom in these cases that they do take and enjoy the antidote so well.

Since on making this visit to us, you have laid aside "the prejudice in favor of taking the body with you," you are able, of course, to visit any number of classes; you can hear the original demonstrations in geometry; the attempts at speech and the successes of the would-be French and German girls, compelled to close attention by having not a word of English to help them in their efforts to understand and be understood; the talk about the last amendment to the Constitution or the aspects of the Indian question, in the class in civil government; and —O "most potent, grave, and reverend" Seniors, forgive us our seeming neglect—the animated discussion on some question in moral science or Butler's Analogy.

In the half-hour after recitations are over, let us go into a choral class, and listen to the noble chords of which we never tire, in such anthems as "I waited for the Lord," or "O rest in the Lord," or to some sweet Ave Maria, or a gay song, equally well rendered in its way. The bell rings for "sections." You remember the first thing in order, and you miss a number of familiar items, if you've been away some years. Changes have been made from year to year which have lessened the number of family regulations, or made some of them requests rather than requirements, but abundant quiet is still secured for study and sleep, and the "still hour," and punctuality everywhere no less insisted on. "Hall exercise" follows, sometimes very short, or omitted entirely if there is an extra engagement for the evening. Before and after supper you will want to pay visits to the reading room and the parlors of whose new furnishings and beauty we wrote you a year ago. We have enjoyed them thoroughly. On Thursday evening, as of old, comes the family gathering for prayer in the seminary hall; on Tuesday evening, a general recess meeting in the lecture room; on other evenings, the little recess meeting—"sweet hour of prayer." How very few of all Holyoke's daughters to whom the name does not recall some of the tenderest and most sacred hours of life, when "Heaven came down our souls to greet." Dear sisters all, "though sundered far," gather often, we pray you, "around one common mercy seat" to ask that the recess meeting and the "half-hour," may be held sacred always, and prized as the most valuable hours of these student days.

The religious history of these years has been similar to that of the period which preceded it. In 1882–3, it was noted that there was "growth in grace among Christians, and marked attention on the part of all the school. Cases of conversion were attended by decided conviction of sin, though the work went on quietly." In 1883–4, the journal says, "As the weeks have gone by, one and another has said for the first time, 'My Saviour!' We have seen with deep delight that many of our girls are steadily growing into 'perfect women, nobly planned'; we know that a quiet work of grace has gone on in many hearts, because of its daily fruits—the prayers in the recess meetings, the faithfulness in work, the Christian spirit shown in many ways."

The year 1884–5 was one long to be remembered, on account of a marked and gracious spiritual quickening early in the first term, the influence of which continued through the year, and perhaps in a degree to the present time. It was in part the fruit of special labors in the parish, in the benefits of which the seminary family shared. Mr. Moody spoke twice in the church, at the outset, and once at the seminary. He was followed by Rev. Rufus Underwood, the evangelist, who, with the pastor, Dr. Love, held meetings in the church for two or three weeks, and frequently spoke at the seminary. "What this season of progress has been to Christians, as well as to others," says the journal, "was apparent in a prayer meeting near the close of the term. Miss Blanchard had invited the young ladies to testify of the spiritual blessings they had received, in notes without signature. It strengthened our faith to hear the words that came in response. They had 'become Christians'; they had 'learned to love God's word'; they had 'received answer to the prayer of years'; they had 'found the half-hour precious'; they had 'felt the love of Christ inspiring them to service.' This has been shown in many little acts of self-denial for Christ's sake, and in the spirit of responsiveness to the many appeals for benevolent objects."

In 1885-6,—"more than half of those who at the opening of the year did not count themselves disciples have given evidence of a change of heart; and some of the few remaining seem not far from the kingdom of God. . . . There has been a prayerful spirit on the part of many, and earnest Christian work."

There is a temperance society to which many of the students belong,—a branch of the Women's Christian Temperance Union,—whose enthusiastic meetings, conducted by themselves, are attended by nearly the whole family. The monthly concert of prayer for missions is regularly observed, as always in the past; and reports on special topics are presented by some of the young ladies, while facts of interest are voluntarily contributed by others. The regular weekly meeting of the school is becoming largely one of conference as well as prayer, helpful thus to better mutual acquaintance, and closer sympathy in the Christian life.

The number and variety of Christian enterprises that in these days come before the school is great, and it has its representatives in nearly all of them. Frequent letters from well remembered friends who are teaching among the freedmen or the poor whites at the South, among the Indians or the Mormons at the West, make the need of such labors seem real, and their reward worth the self-sacrifice they cost. The intimate connection with the Holyoke schools in Cape Colony—which in their spirit and aims are so truly in sympathy with missionary work, and are already doing so much in that line—tends to a like result.

During Miss Blanchard's administration, several of the seminary teachers have gone to take up some form of Christian service in foreign lands. Two of these are in Turkey: Miss Helen E. Melvin, of the class of '79, who went in 1883 to teach at Constantinople; and Miss Ella T. Bray, of the class of '83,—now Mrs. Dr. Graham of Aintab,—who went out in 1885. Miss Mary Ella Spooner, of the class of '72, and a teacher at the seminary from that time, left in 1884, to teach in the Hawaiian Islands, as lady principal of the Oahu College; and in 1887, Mrs. Sophie (Smith) Burt, a graduate of Oberlin, and a seminary teacher for two years, went also to the islands, to assist in the school for native boys at Hilo, of which her husband, Rev. Arthur W. Burt, is principal.

In June, 1884, the question of appointing women on the board of trustees—which had been for some time under consideration—was affirmatively settled. It was voted that henceforth the principal should be a member *ex officio*, and Mrs. A. Lyman Williston also was chosen a trustee. In 1886, Mrs. Helen (French) Gulliver was elected. At the same time, it was decided that the principal should be *ex officio* a member of the executive committee.

In 1885, the steward, Mr. Ithiel Lawrence, resigned the position which he had ably filled since 1866. The board passed a vote highly commending the efficiency and fidelity with which Mr. Lawrence had conducted the business committed to his care, and expressing "the most hearty good wishes for his welfare, and that of his highly esteemed wife, during their remaining years." Mr. Lawrence was succeeded by Mr. David E. Phillips, of Columbus, Ohio, who filled the position with great ability and acceptance; but resigned in 1886 on account of business inducements elsewhere. The present steward is Mr. Lewis H. Porter, of Williamsburgh, Massachusetts.

In concluding the history of this period, it is proper that there should be a brief statement of the present financial condition of the seminary. Its principal permanent funds are, that for aiding deserving students, and the "general fund," the income of which is at the disposal of the board for the various needs of the institution.

The need of an education fund was early recognized, and efforts were made by trustees to secure donations for this object; but, previous to 1872, the amount obtained was only five thousand dollars. A legacy of fifteen thousand dollars expressly designed for the education fund, from Miss Phebe Hazeltine, of Boscawen, New Hampshire, was received in 1872-3; and other legacies and donations which came in during Miss Ward's administration raised the total amount to twenty-eight thousand dollars. Since that time, it has increased to forty-five thousand dollars. There are, besides, two loan funds of five thousand dollars each, for the same purpose; one of which was given in 1882, by Homer Merriam, Esq., of Springfield, and the other in 1884, by Edward Smith, Esq., of Enfield, Massachusetts. The income of these, by request of the donors, is loaned, rather than given, "in order to promote self-reliance and self-respect on the part of the pupil," who gives her note and pays interest. The gentleman last named, in addition to his own donation, labored much in soliciting gifts for the education fund from others; and the board passed a vote of thanks for his assiduous efforts.

The general fund has been very small until recently. Many years ago, a legacy of five thousand dollars was received from Mr. Henry Kendall, of Leominster, Massachusetts, the income of which has been used for general expenses. It was stated that he was led to make the bequest by observing the benefit received at the seminary by a certain pupil. In 1883, a legacy of twenty thousand dollars from Mr. John B. Eldridge, of Hartford, was added to the general fund; and in 1885, a like amount—the larger part of a bequest from Mr. Eber Gridley, of Hartford—was appropriated to the same purpose, the remainder being reserved for buildings and improvements. About five thousand dollars from other bequests has been added to the fund.

Some years since, Mrs. Julia M. Tolman, formerly associate principal, left to the seminary between three and four thousand dollars, the income of which should be used for special grants to teachers, for purposes of health and improvement. It was the hope of the donor that this might be the beginning of a much larger provision for the object. Small as the fund is, it has been of inestimable service to the teachers, on the one hand by furnishing them the means of rest and recreation from time to time; and on the other, by enabling them to enlarge their resources by attending lectures, or visiting other institutions of learning.

The following table recapitulates the amounts given above, including also the Goodnow Park fund and the Boswell fund for the library, previously mentioned:—

For aiding students,			
	Education Fund,	$45,000 00	
	Homer Merriam Loan Fund,	5,000 00	
	Edward Smith Loan Fund,	5,000 00	$55,000 00
For general purposes,			
	Eldridge Bequest,	$20,000 00	
	Gridley Bequest,	20,000 00	
	Other Bequests,	10,000 00	50,000 00
Tolman Fund,			3,610 00
Goodnow Park Fund,			5,000 00
Boswell Fund,			1,000 00

Thus far, the seminary has done its work with little income besides the receipts for board and tuition of pupils. It has almost invariably succeeded in "making the ends meet," year by year, so far as ordinary expenses are concerned; often with a slight surplus, which has been carefully saved toward the special outlays sometimes required. There is now in progress an effort by the alumnæ to raise the modest sum of twenty thousand dollars, to be called "The Mary Lyon Fund," for endowing the principal's chair. Endowments for all the departments are no less essential, in order that the continued growth and prosperity of the institution —so far as they depend on material resources—may be reasonably assured.

Compared with earlier years a much larger proportion of the young ladies enter with the intention of completing the course, and actually do so. Candidates are not admitted until they are sixteen years of age, and many are older. A recent junior class on entering averaged eighteen years and two months. The age at graduation is generally between twenty-one and twenty-two.

Requirements for admission now include, besides the common English branches, preparation in Latin, algebra, and geometry. The fiftieth annual catalogue contains the following

SYNOPSIS OF THE COURSE OF STUDY:

JUNIOR YEAR.

Latin: Cicero de Senectute; with exercises in Prose Composition, and in Reading at Sight. *Mathematics:* Olney's University Algebra and Chauvenet's Geometry. *History:* Ancient, to the Fall of the Roman Empire. *Constitution of the United States:* Andrew's Manual. *Physiology:* Huxley and Youmans. *Botany:* General Morphology, and Classification of Phænogams. *Rhetoric,* with Exercises in English Composition. *Elocution:* Instruction in Classes. *Vocal Music:* Lessons in Choral Classes. *Gymnastics.*

JUNIOR MIDDLE YEAR.

French or German: Text-books of Sauveur and Wenckebach. *Mathematics:* Olney's Trigonometry. *History:* Mediæval and Modern. *Botany* General Histology and Physiology; Systematic Botany continued. *Mineralogy:* Dana's Manual. *Zoölogy:* Packard's; Jordan's Manual of Vertebrates. *Chemistry:* Houston's, with Eliot and Storer's Manual. *Elective:* Biology, Advanced Chemistry, or Mathematics. *Elocution:* Class Exercises. *Vocal Music:* Choral Classes. *English Composition. Gymnastics.*

SENIOR MIDDLE YEAR.

Latin: Virgil's Æneid. *Physics:* Atkinson's Ganot. *Astronomy:* Newcomb and Holden's. *Geology:* Dana's. *English Literature:* Shaw's Manual and Brooke's Primer, with topical studies of authors. *Rhetoric:* Welsh's Complete Rhetoric. *Elective:* Livy or Tacitus; advanced work in Natural Science, Literature, or Mathematics. *Elocution:* Class Exercises. *English Composition. Vocal Music:* Choral Classes. *Gymnastics.*

SENIOR YEAR.

Latin: Cicero de Immortalitate; Selections from Horace. *Ancient Literature:* Outlines from Quackenbos and Schlegel, with topical studies of authors. *Psychology:* Hickok's. *Ethics:* Hickok's Moral Science. *History of Art:* Outlines from Lübke, and topical studies. *Theism and Christian Evidences:* Studies from Fisher's Grounds of Theistic and Christian Belief, Wright's Logic of Christian Evidences, and Butler's Analogy. *English Composition. Elocution:* Class Exercises. *Vocal Music:* Choral Classes. *Gymnastics.*

COURSE OF BIBLE STUDY.

A comprehensive course of Bible study is regularly pursued, in weekly lessons, throughout the four years, as follows:—

Junior Year—Genesis and Exodus; the Gospels. *Junior Middle Year*—Joshua, Judges, and Samuel; Acts. *Senior Middle Year*—Kings and Chronicles; Hebrews. *Senior Year*—The Prophetical Books; Romans.

OPTIONAL COURSES IN LANGUAGES.

The following optional courses are pursued in Greek, French, and German: To one who completes either, a special diploma is given, as they are additional to the regular curriculum.

GREEK COURSE.

First Year—White's Lessons and Goodwin's Grammar; Xenophon's Anabasis. *Second Year*—Selections from Herodotus and Thucydides; Homer's Iliad or Odyssey. *Third Year*—Demosthenes's Olynthiacs and Philippics; Prometheus of Æschylus. *Fourth Year*—Plato's Apology; Antigone of Sophocles; Alcestis of Euripides. Prose Composition and reading at sight, each year.

FRENCH COURSE.

First Year—Sauveur's Causeries avec mes Élèves; Fables de La Fontaine, some of them committed to memory; Grammar. *Second Year*—Sauveur's Contes Merveilleux, read and made the subject of conversation; Littérature Française Contemporaine, by Pylodet; Grammaire

Française pour les Anglais, by Sauveur; Written and Oral Exercises. *Third Year*—Sauveur's Grammar concluded; Select Modern Plays; Poetry; Athalie by Racine. *Fourth Year*—Histoire de la Littérature Française, by Cart; Translations from English into French; Compositions; L'Avare, by Molière; Le Cid, by Corneille; Selections from Madame de Sévigné, La Bruyère, Mérimée, Victor Hugo, George Sand, Lamartine.

GERMAN COURSE.

First Year—Object Lessons: Deutscher Anschauungs-Unterricht für Amerikaner, by Wenckebach; Das Deutsche Buch der Sauveur Schule; Grammar; Poetry committed to memory. *Second Year*—Grammar; Written and Oral Exercises; Deutsche Grammatik für Amerikaner, by Wenckebach; Grimm's Märchen; Höher als die Kirche, by Wilh. von Hillern; Undine, by Fouqué; Poetry: Die Schönsten Deutschen Lieder, selected by Wenckebach. *Third Year*—Grammar; German Prose Composition, by Buchheim; Lyric Poems; Die Jungfrau von Orleans, by Schiller; Minna von Barnhelm, by Lessing. *Fourth Year*—History of German Literature; Letters and Compositions; Poetry: Ballads by Goethe, Schiller, and Uhland; Schiller's Wilhelm Tell; Lessing's Nathan der Weise; Goethe's Iphigenie auf Tauris.

DRAWING AND PAINTING.

Lessons are taken in charcoal-drawing from casts or models, and in sketching from nature. A normal class, also, is instructed in the elementary forms of design, and in outline drawing. Painting is taught, in both water-colors and oils.

MUSIC.

All the students have regular lessons in choral classes. Private instruction in the cultivation of the voice, and in piano practice, is given by teachers who have studied in conservatories in this country and in Germany.

CHAPTER XVIII.

STATISTICS.

THE following statistics relating to the number of teachers and students, with expenses, year by year, for the half century, are compiled from the annual catalogues of the seminary.

Before 1862 the course of study occupied three years; since then, four years. The late senior classes would have been larger had not many taken time for extra study in different departments before, or instead of, finishing the course.

Since 1876 the charge for board and tuition has included lights, steam heating, and lectures.

Years.	Teachers.	Assistant Pupils.	CLASS. Senior.	Middle.	Junior.	Whole No. Students.	Board and Tuition.
1838	4	3	4	31	78	116	$64
1839	5	3	12	31	60	103	60
1840	6	4	17	40	62	119	60
1841	5	3	10	27	70	113	60
1842	8	3	15	50	107	172	60
1843	13	2	16	50	118	184	60
1844	12	1	34	66	106	206	60
1845	18	2	51	72	123	246	60
1846	16	1	42	69	71	182	60
1847	13	2	44	59	85	188	60
1848	12	2	47	62	126	235	60
1849	13	2	23	58	138	219	60
1850	16	0	34	69	121	224	60
1851	13	2	60	55	129	244	60
1852	16	2	31	59	162	252	60
1853	16	1	46	49	163	258	60
1854	17	2	43	55	180	278	68
1855	19	1	57	73	162	292	68
1856	19	2	49	70	156	275	75
1857	18	0	59	68	139	266	80
1858	20	1	57	71	147	275	80
1859	19	1	56	70	150	276	80
1860	22	1	42	70	148	260	80
1861	22	1	66	65	157	288	80

STATISTICS.

Years.	Teachers.	Asst. Pupils.	Senior.	Sr. Middle.	Jr. Middle.	Junior.	Whole No. Students.	Board and Tuition.
1862	23	1	56	43	70	85	254	$80
1863	23	0	40	47	87	127	301	80
1864	23	1	51	41	98	153	343	80
1865	23	1	38	47	87	117	289	125
1866	24	1	60	40	87	100	287	125
1867	24	0	59	38	59	124	280	150
1868	24	0	45	31	64	122	262	150
1869	25	0	38	30	64	136	268	150
1870	26	0	33	40	84	111	268	150
1871	27	0	37	34	77	132	280	150
1872	29	0	42	40	74	118	274	150
1873	28	0	48	40	75	108	271	150
1874	30	1	37	42	74	148	301	150
1875	27	1	29	36	95	128	288	150
1876	27	0	39	55	77	113	283	175
1877	28	0	44	31	73	114	262	175
1878	27	0	29	41	84	97	251	175
1879	28	1	31	38	86	118	273	175
1880	27	1	33	48	68	77	226	175
1881	26	1	47	35	84	83	249	175
1882	28	0	30	29	90	118	267	175
1883	28	0	42	45	81	121	289	175
1884	29	1	47	43	78	119	287	175
1885	30	0	27	56	70	116	269	175
1886	30	0	53	48	73	120	294	175
1887	33	0	47	52	65	149	313	175

Average number of teachers for the fifty years, . . . 20.78
Average number of assistant pupils, ... 1.04
Average charge, ... $110.30
Number of different students, about 6,360

	50 Years.	Last 25 Years.	Last 5 Years.
Whole number of names,	12,500	6,975	1,452
Average " students,	249.88	278.96	290.2
Whole " graduates,	1,997	1,026	216
Average " graduates,	39.94	41.04	43.2
" middle class, first 24 years, 58.			
" senior middle class,		40.96	48.4
" junior middle class,		78.28	73.4
" junior class,	120.32	118.68	125.2
Class average,	73	69.74	72.55

The following table shows the attendance, by years, from different states and territories of the Union, and from other countries. The increase of totals over those just given is due to the added names of resident graduates.

MOUNT HOLYOKE SEMINARY.

A Table Showing the Attendance from Different Places Each Year.

Years.	Alabama.	Arkansas.	California.	Colorado.	Connecticut.	Delaware.	District of Columbia.	Florida.	Georgia.	Idaho.	Illinois.	Indiana.	Indian Territory.	Iowa.	Kansas.	Kentucky.	Louisiana.	Maine.	Maryland.	Massachusetts.	Michigan.	Minnesota.	Missouri.	Mississippi.	Nebraska.	Nevada.	New Hampshire.	New Jersey.
1838	.	.	.	.	13	.	.	.	.	.	.	.	.	.	.	.	.	1	.	82	.	.	.	.	.	.	3	**3**
1839	.	.	.	.	13	.	.	.	.	.	.	.	.	.	.	.	.	1	.	64	.	.	1	.	.	.	6	**2**
1840	.	.	.	.	24	.	.	.	.	.	.	.	.	.	.	.	.	.	.	68	.	.	.	2	.	.	7	**2**
1841	.	.	.	.	18	.	.	.	.	.	.	.	.	.	.	.	.	.	.	64	.	.	.	1	.	.	11	**1**
1842	.	.	.	.	36	1	.	.	.	.	.	.	.	.	.	.	.	.	.	87	1	.	.	1	.	.	14	**6**
1843	.	.	.	.	31	.	.	2	.	.	.	.	.	1	.	.	.	.	.	89	.	.	.	.	.	.	13	**7**
1844	.	.	.	.	41	.	.	.	.	.	1	1	.	.	.	.	.	.	.	93	.	.	.	.	.	.	11	**4**
1845	1	.	.	.	45	.	.	.	.	.	1	1	.	.	.	.	.	.	.	109	.	.	.	.	.	.	18	**3**
1846	.	.	.	.	25	.	.	.	.	.	.	1	.	.	.	.	.	3	.	78	.	.	.	.	.	.	15	**4**
1847	.	.	.	.	32	1	.	.	.	.	.	.	1	.	.	.	.	3	.	71	.	.	.	.	.	.	14	**5**
1848	.	.	.	.	44	.	.	.	.	.	2	.	1	1	1	.	.	12	.	79	.	.	1	.	.	.	25	**6**
1849	.	.	.	.	32	.	.	.	.	.	.	1	2	.	.	.	1	12	1	75	2	.	.	.	.	.	25	**8**
1850	.	.	.	.	33	.	.	.	.	.	2	.	2	.	.	.	.	7	.	78	3	.	.	.	.	.	20	**8**
1851	.	.	.	.	49	.	.	.	.	.	6	.	.	.	.	.	.	1	.	73	4	.	.	.	.	.	22	**8**
1852	.	.	.	.	32	.	.	1	.	.	4	.	1	1	.	.	.	9	.	77	1	.	.	1	.	.	16	**14**
1853	.	.	.	.	45	.	.	.	.	.	4	.	2	.	.	.	.	5	.	77	.	.	.	1	.	.	22	**20**
1854	.	.	.	.	52	1	.	.	.	.	9	2	3	.	.	.	.	7	.	68	1	.	1	.	.	.	30	**16**
1855	.	2	.	.	42	1	.	.	.	.	6	5	.	.	.	.	.	15	.	91	3	.	.	.	.	.	24	**9**
1856	.	.	.	.	42	.	.	.	.	.	7	3	.	.	1	.	.	21	.	75	.	1	1	.	.	.	25	**6**
1857	1	.	.	.	35	1	.	.	1	.	5	4	.	1	1	.	.	23	.	75	.	1	.	.	.	.	30	**13**
1858	.	.	.	.	33	.	.	.	1	.	7	5	.	1	1	.	.	12	.	96	3	.	.	.	.	.	24	**7**
1859	.	.	.	.	39	1	.	.	1	.	9	1	.	1	1	.	.	14	.	91	2	.	.	.	.	.	22	**3**
1860	.	.	.	.	41	1	.	1	1	.	7	3	.	.	.	.	.	16	.	77	4	.	1	.	.	.	23	**4**
1861	.	.	.	.	40	.	1	.	.	.	1	2	.	1	.	1	.	15	.	90	4	.	1	.	.	.	21	**9**
1862	.	.	.	.	32	.	1	.	.	.	3	3	.	1	.	1	.	10	.	83	4	.	.	1	.	.	19	**8**
1863	.	.	.	.	32	.	2	.	.	.	5	3	.	.	.	3	.	17	.	101	5	.	1	.	.	.	25	**14**
1864	.	.	.	.	40	.	.	.	.	.	6	3	.	1	.	2	.	19	.	115	5	.	1	1	.	.	23	**25**
1865	.	.	.	2	44	1	1	.	.	.	9	.	.	.	.	1	.	11	.	88	5	.	.	.	.	.	16	**21**
1866	.	.	.	1	41	1	.	.	.	.	4	1	.	.	.	1	.	14	.	85	3	.	.	.	.	.	21	**15**
1867	.	.	.	3	37	.	.	.	.	.	6	.	.	1	.	.	.	14	.	100	1	.	1	.	.	.	20	**12**
1868	1	.	.	.	32	1	.	.	.	.	3	.	.	1	.	.	.	22	1	85	.	1	1	.	.	.	23	**8**
1869	.	.	.	.	44	.	.	1	.	.	1	1	.	3	1	.	.	15	.	90	.	.	.	2	.	.	20	**12**
1870	.	.	.	.	35	1	.	1	.	.	1	2	.	5	.	.	.	14	.	98	1	.	1	1	.	.	16	**8**
1871	.	.	1	.	43	1	1	.	.	.	2	1	.	1	1	.	.	13	.	102	.	1	.	.	.	.	18	**8**
1872	.	.	1	1	48	1	1	.	.	.	1	7	.	.	.	.	.	8	.	87	1	1	1	.	.	.	13	**13**
1873	.	.	.	.	46	1	1	.	.	.	3	2	.	.	.	.	.	11	.	95	3	1	.	.	.	1	11	**10**
1874	.	.	.	.	54	1	2	.	.	.	4	2	.	1	1	.	.	9	1	102	4	2	.	.	.	1	14	**11**
1875	.	.	.	.	65	.	2	1	.	.	3	2	.	.	.	.	.	9	1	101	1	.	.	.	.	.	14	**12**
1876	.	.	.	.	55	.	1	1	1	.	5	3	.	.	.	1	1	13	2	86	1	2	.	.	.	.	10	**9**
1877	.	.	.	.	46	2	2	.	.	.	7	2	.	.	1	1	1	5	1	74	1	3	4	.	1	1	13	**8**
1878	.	.	.	.	5	2	1	.	.	.	7	1	.	1	1	.	.	1	.	67	2	2	3	.	1	.	14	**13**
1879	.	.	.	.	56	1	2	.	.	.	7	.	.	.	1	.	.	5	2	73	3	1	1	.	1	.	10	**13**
1880	.	.	.	.	43	1	1	.	.	.	7	1	.	3	1	.	.	4	.	59	2	2	2	.	.	.	5	**11**
1881	.	.	.	.	42	.	1	.	.	.	9	3	.	3	1	.	.	6	1	62	3	3	1	.	.	.	12	**10**
1882	.	.	1	.	40	1	1	.	.	.	7	1	.	8	1	.	1	6	2	68	.	3	.	1	.	.	9	**17**
1883	.	.	1	1	46	.	2	.	.	.	5	1	.	11	1	.	1	6	1	73	.	7	.	1	.	.	8	**14**
1884	.	.	.	1	48	1	1	.	.	.	4	1	.	8	.	.	.	6	3	85	1	3	1	.	.	.	16	**13**
1885	.	.	.	1	31	2	1	.	1	.	7	1	.	7	.	.	.	9	3	86	.	3	.	.	.	.	13	**14**
1886	.	.	.	2	39	.	2	.	.	.	14	1	.	2	.	.	1	12	.	92	.	2	3	.	2	.	12	**16**
1887	.	.	1	1	46	.	1	.	.	.	17	.	.	12	3	.	1	16	.	87	3	.	2	.	2	.	9	**11**

STATISTICS.

A Table Showing the Attendance from Different Places Each Year.

Years	United States and Territories													Other Countries															Total
	New York	North Carolina	Ohio	Pennsylvania	Rhode Island	South Carolina	Tennessee	Texas	Vermont	Virginia	Washington Territory	West Virginia	Wisconsin	Canada	New Brunswick	Nova Scotia	Bermuda	Brazil	Cape Colony	China	Hawaiian Islands	Holland	India	Italy	Natal	Persia	Syria	Turkey	West Indies	
1838	3	1	1	3					7																					116
1839	8								8																					103
1840	3	2							11																					119
1841	10								8																					113
1842	16	1		1					6				1								1									172
1843	29	1		1					7	1		1	1																	184
1844	32	1	2						18	1											1								1	206
1845	42	4	1						19												1								1	246
1846	39	4	1						9			2																		182
1847	36	3	1						19			2																		188
1848	33	4	3	1					16	1		3	1																	235
1849	34	4	4	2					10	1		2	1	1							1									219
1850	36	2	5	1					18			6	1	1							1									224
1851	34	6	8	1					24			5									3									244
1852	44	6	7	1					26	2		1	1								6	1								252
1853	40	1	8	7	2				13	2		2	2								4	1								258
1854	44		9	6	2		1		20	1		2				1					1	1								278
1855	53		8	8	2		1		16			1		3								2								292
1856	53		4	10	3				12	1		3		5								1							1	275
1857	42		3	9	1				11			1		5								2							1	266
1858	40		12	11	4				17			1																		275
1859	44		16	6	3		1	1	15			2										2							1	276
1860	42		8	9	4		1		13			1		1																260
1861	48		6	6	3				26			2		1	1															288
1862	39		5	16	2				19	1		1		1	2							1				1				254
1863	55		3	9	3		1		11			3	1	3								1		2		1			1	301
1864	56		4	10	3		1		19			4							1			2		1		1			1	343
1865	48		4	7	2				16		1	7										3							1	289
1866	40	1	11	11	4		1		15		1	9								2		2						1	1	287
1867	30	1	5	11	6				18			8	1	2	1			1				1						1		280
1868	44		4	7	1		1		17			3	1	2							1							1		262
1869	41		5	3	4		1		16			2	2	1					1		1									268
1870	35		2	10	4				25			1	1						1			4							1	268
1871	43	6	9	4					20			1	1									2								**280**
1872	34		2	11	7				29			1	1								1	4								**274**
1873	36		3	5	5	1			30			1	1	1					2									1		**271**
1874	35		2	7	5	1			34			1	2						2		1	1					1		**301**
1875	33		4	7	6				21			1	1								1	2								**288**
1876	30		9	10	5		1	1	24	1			1									2						1		**282**
1877	35		7	11	3				23			1	1								1	3						3		263
1878	49		6	11	2				14				1								1	1						1		247
1879	45		11	12	2			3	16			3									1	4								273
1880	37		12	9	1	1			15		1	5		1								1								225
1881	44		8	16	2	1		1	14	1		2	3	1								1								251
1882	39		10	21	1			1	19		1	3	3		2		1								1					268
1883	44		9	21	1	1		1	22		1	3	1	2			1							1						289
1884	44		5	25		1		1	14	1	1			2			1				1		1	1						289
1885	42	3	4	24		1			10	2		1		1			1				1		1	1						270
1886	49	2	3	20	2		1	1	14	1											1	1	1					1		297
1887	52	2	7	22	1		1		11	1		2									1	1						2		315

CHAPTER XIX.

RESULTS: TESTIMONY OF ALUMNÆ.

IT is impossible in our day to trace fully the influence of the seminary, but certain results are already manifest; and though it is not desirable to count up the fruits of human labor, it is due to the praise of divine power to speak of what God has done through the seminary. Many observers have borne witness to the mental and moral efficiency of those it has trained for Christ and the church; and parents are constantly expressing their gratitude for its work; but the present record allows room for more personal statements only. The testimony of alumnæ alone, whose experience in life has tested the value of their Holyoke training, would fill many chapters. The following selections from replies to a few questions in a recent circular represent all the past years of the half century, and the date with each indicates the time of the writer's connection with the seminary.

The reader will pardon the repetition inseparable from the testimony of so many to the same points.

"DEAR ALMA MATER: The story of your life will scarcely be told in four hundred pages if a tithe of what your daughters might say should be woven into the web.

"The certificate that gives me the right to call you Alma Mater bears date August 5, 1852. You ask what proves to be the value of my Holyoke education, and whether it has helped me 'to be, or do, or bear.' These are rather personal questions, dear Alma Mater. I should fear to give very positive answers concerning myself. But I think Dr. Wilson was not far wrong, if he said, as reported recently in New York City, that the Holyoke graduates were noted for 'Christian culture and common sense.'

"Many of your 'ways,' dear Mountain Mother, used to seem to me unnecessary. To tell the truth, I sometimes thought you very fussy. But now I justify them nearly all, and some of them I almost glorify. Let me mention the silent and solitary 'half-hour.' How could you fit your daughters to bear this rushing life of the nineteenth century, without teaching them the blessedness of being alone with God? Some souls learn its sweetness more readily than others, but with all it is a matter of education, and I am sure that the mature life of every woman will justify the means by which she was led 'Nearer, my God, to thee!'"

'55. "I am too busy with present duties to get much time for retrospect."

'46. "But it would be a guilty silence to withhold my humble testimony to the excellence of my Alma Mater, in whose halls I spent some of the best and happiest years of my life."

'47. "The influence of the seminary has been so interwoven with my life that only eternity can reveal even to myself how much I owe it."

'50. "It has been to me these thirty-five years a comfort, joy, and help."

'70. "I would not have it taken out of my life for any imaginable exchange."

'42. "That life is to be consecrated to the renovation of a fallen world, not frittered away in selfish indulgence, was the very spirit of the place."

'51. "The seminary trained the conscience to carefulness in little things. I cannot well imagine a graduate taking a place among so-called 'society ladies,' or giving her time to the round of fashion. We naturally expect her to enter on the work of doing good, carrying out the principles taught by Mary Lyon."

'67. "The influence of seminary teaching unfitted me for enjoying fashionable society."

'65. "It fits women for almost every sphere except that of the devotee of fashion."

'65. "I am not a butterfly—will not despise butterflies —but I must be of some practical use in the world."

'55. "To me, after thirty years, Mount Holyoke Seminary has a sacred sweetness, a special unworldliness about it, that is near akin to the Palace Beautiful, and the Delectable Mountains of Bunyan. I don't mean that everything was always pleasant and easy. I remember when I thought myself near dying with the most painful disease to which the young are ever liable, *nostalgia*. Such multitudes of strange people surging into examination rooms! I believe no one was so homesick as I, going into all public places where duty called, with aching heart and streaming eyes! As for the domestic work, my memory brings up only one impression,—I thought it 'lots of fun'! I used to drive dull care away as well as algebra and Nepos, and laugh and joke, leading the white crockery circle, setting up china, or cleaning silver. The thoroughness of the teaching was something that I always appreciated. There is a self-respect in doing a thing well and not shirking. A building, high or low, is better for a good foundation; and who knows what the Master Architect may add at any time? When I came to teach in Tennessee, at Andover, and in Persia, I was glad of everything I had learned, and I found that the principles I had been taught were the only ones to give satisfaction. I am impressed by the value of the Mary Lyon plans; they are as valuable in Persia as in Massachusetts.

"I never can forget the impression of awe awakened by the very atmosphere at the seminary. I went there right from the world and found it a place peculiar and apart. At morning devotions in the hall, Miss Spofford spoke of God; he seemed so glorious and real that my homesick heart leaned hard on him as a heavenly Father. On the way down to breakfast the first Sabbath, a teacher took my hand lovingly just a moment on the stairs and said, 'I hope that the Saviour may be so near that you will never forget your first Sabbath here.' I never did. The 'silent time' was an unspeakable means of grace. The system of the school was throughout an appeal to conscience. No spies—no watching—the reins of government placed in our own hands with only God as witness to our truthfulness; bound by such chains of honor and duty who could deceive? Did the temptation come, 'Do this, no one will know it,' it was sufficient to remember, 'Myself would know it.' In one sense it was freedom uncontrolled, in another we were held to do right as the world obeys its own law of gravitation. Missionaries and Christian workers were born there. After such a stand, it was not easy to go back even when the Master said, 'Go ye into all the world.' So Mount Holyoke's light has streamed afar, and its lines are gone out through all the earth."

'49. "What has my Holyoke education helped me to be, or do, or bear? I incline to answer 'Everything.' My Christian life began in the seminary. There I learned to hold myself responsible for my part of the Christian work in the world, just as for my daily share of the domestic work; I learned to take part in and to conduct meetings, and to try to win others to Christ as I myself was won—by personal appeals in the friendly way which is habitual there. During the three years spent in teaching after graduation I learned that the influences of the seminary are not confined within its walls but may be carried into other schools with gracious results. Whatever I have been permitted to do as a teacher, a minister's wife, or a mother, I have been taught to do as unto Christ."

'47. "I have sometimes been constrained to say that everything that is good in me is due to the seminary; whereupon my husband asks, 'And is everything that is bad in you due to me?'"

'38. "In the first year of the seminary I was taught self-denial for the good of others, and to do with my might what my hands found to do. I have been teaching ever since, though now nearly seventy years of age. For fifteen years before the war I taught in Virginia. I have been superintendent and teacher in a union Sabbath-school in St. Augustine, Florida, for twenty-five years. Three years I taught the Indian prisoners in the fort. My sixtieth birthday I celebrated by accompanying Captain Pratt to Dakota to collect Indians for the Carlisle school. We brought eighty-seven boys and girls. The same fall we went to Indian Territory and brought fifty more; with these one hundred and thirty-seven Indians, Carlisle school was opened. The following year I went to Dakota again with Captain Pratt, and brought back both boys and girls for his school. In 1883, after great efforts, we succeeded in St. Augustine in establishing a school for colored girls, giving them a three years' course in English studies. This may seem egotistical, but Miss Lyon is working through her pupils not only in foreign lands but in many places in our own country. Let her have the credit due."

'48. "I cannot hope that what I write will be of the least worth, save as an added testimony—a reference to what Holyoke has led me to wish and strive to do, rather than to what has been actually done."

'56. "In every position I have occupied, my Holyoke training has been a help and a blessing; no part of it was useless, not even the 'side-step' in calisthenics; and as the years roll on, I am more and more grateful for it. The lessons of promptness and faithfulness in every duty left an indelible impression. The old motto on our domestic work books—'Be prompt, be faithful, be diligent'—has been to many of us a constant help and incentive. No one who rang the bells could fail to realize the importance of using every moment to the best advantage, or to learn that every second had its value, whose loss was irreparable. Accepting for three years irksome and sometimes unaccustomed domestic work, not as a disagreeable necessity—but pleasantly—

because it was a plain duty, the habit became confirmed, and made it easier to meet the discomforts of after years in the same spirit of cheerful acquiescence. In these days of steam heat and electric bells, the old 'rising bell' should have an honored place among Franklin stoves, wood-baskets, and other relics of the early days—specially honored because its mission has been accomplished. Peace to its clapper, evermore!"

'43. "To bring out all the powers, moral, mental, and physical, in perfect harmony, was the problem. To aid in this the domestic department was invaluable. The tact, skill, and good judgment required is educational in itself, giving self-reliance and self-respect. Its influence helps in after life to maintain an orderly home and thus throws a charm over home life."

'63. "I distinctly recall Miss Jessup's saying: 'Young ladies, in order that all may run smoothly, everything must be done correctly and in time. You who have charge of meals should be particularly prompt. If they are delayed ten minutes, or even one, multiply that by the number you have kept waiting, and see how many minutes are lost. Always have in your room a book that can be read at intervals, or a piece of work ready to pick up for a few moments, and you will be surprised at the end of a month to see how much you have accomplished.' These remarks made such an impression on me, that the habit then formed has followed me through life. Now that I have family cares, much is accomplished in this way that otherwise had been impossible."

'67. "To me it was a constant object lesson to see those many domestic circles revolve in time and tune because somebody had thought it all out; and to see how the comfort of all was secured by each one's doing her part in due season."

'83. "The domestic cares have made me more systematic in all I undertake."

'70. "The grand truth was impressed so that it cannot be forgotten, that good, honest work with the hands is not degrading; that gold is gold whether in the kitchen or in society."

'49. "It is much to have borne a part in household cares at the seminary. It was a daily object lesson in system and order—a beautiful example of successful co-operative housekeeping."

'83. "When my friends were considering the question of sending me to Holyoke, the domestic work was an objection; but the household duties never burdened me. Many happy hours were spent in the domestic hall, and many firm friends made on the different circles. It was sometimes a sacrifice to be aroused from slumber early in the morning to work on the 'breakfast circle'; but every sacrifice helps us to make greater ones."

'54. "It taught us how to share responsibility with others."

'39. "The domestic arrangements assisted much in developing character, and forming habits of neatness, promptness, system, and consideration for others. It taught us not to despise work, nor ignore workers; and made practical the proverb, 'What is worth doing, is worth doing well.'"

'56. "Some of us spend many hours of our lives in household labors; it was well to learn in our school days, that such work is worthy of a cultivated woman. The true Holyoke spirit finds poetry even in housework."

'58. "It is worth something in these days to have had the idea fully ingrained in youth, that an educated woman can attend to her own housework, without injury to her dignity."

'63. "With much opportunity for observation I believe that no other school presents any plan which can take the place of this. In an age when there are so many adverse influences to keep our daughters from acquiring ability to perform for themselves or teach others to perform the duties of the home, when so many unwise mothers look upon service as menial, it has been the glory of Mount Holyoke Seminary to teach the dignity of labor."

'59. "The share in the household cares helps to maintain the 'sweet home' atmosphere of the place, which is a remarkable feature in so large a school. The cultivation of the spirit of helpfulness, the repression of caste feeling, the more thorough acquaintance of students with each other, and the diversion that comes from entire change of thought, are among the incidental advantages. It tends to develop the feeling of personal responsibility, for it shows plainly that the faithfulness of each is necessary to the good of all."

'55. "Some of my classmates would applaud, if they knew the obligation I now feel to the domestic department. They have listened to my raptures over Miss Jessup's morning lectures on theology and been amazed at my lack of appreciation of her running commentary on the 'weekly items.' The elucidations of theology are forgotten, but I have never in thirty years used a flat-iron, without remembering not to rest it on the ironing sheet. I protect matches with the utmost care, and if I take thick peelings from apples, it is because I do not live up to my instructions. . . . The text-book attainments are worth to me all that I expected. The lectures and the criticisms on composition are invaluable. But now I have grown rusty in these departments; yet all my life, ordinary and extraordinary, is made happier and worthier, from my training in domestic work, care of my room, wardrobe, and accounts. The living by bells, which often seemed so tiresome, cultivated such habits of system that I have been able to accomplish more than I ever expected. Of course, both in family and parish, any system has to be elastic; and I long ago learned not to consider myself interrupted. . . . The training in religious activities is another blessing for which I thank the seminary. The little praying circles were educators not only in spiritual life but in Christian fellowship and conquest of self. Many of us who have since been called to lead in Christian work have thanked our teachers for the perseverance with which they drew out our small abilities in the recess meetings and taught us the preciousness of united prayer. I am ashamed that I was not more enthusiastic in such mission enterprises as were then on foot—like basting patchwork for mission schools; but no time was ever more worthily spent on composition than the eight hours I devoted to the old volumes of the *Missionary Herald* in following the history of one missionary in Africa.

"Miss Catharine McKeen once said, 'The feature which distinguishes Mount Holyoke Seminary from all other schools is the cultivation of the conscience.' Many of us used to grumble at the frequent use of the word 'discipline,' but we are indebted to it for much that makes us of service in the world. The impression left on the pupils is, that life is to be used for God; that we are personally responsible for all that we can do, not for what we would do. Holyoke girls are ambitious, not to shine or be admired, but to be of use in the world. We were so constantly taught that we were not our own that it was less difficult to yield the will of one for the good of the many."

'55. "We learned that if we would have a sound mind in a sound body we must mingle active with sedentary employments and never fall into the absurd notion that in order to be refined ladies we must be indolent, or become those inefficient, fainting, nervous things that sometimes pass under that name."

'44. "No one can attend that school and not learn that living is not merely existence. It makes women of practical common sense. No one can be a pretender there. Drones will not stay. They will be sick and leave. To bear a part in household cares avails much in making systematic, neat, economical housekeepers; economical of time as well as means. My one year at Holyoke has helped me to plan and control a school of five hundred pupils; to cater for a large household;

to systematize and reduce manual labor; and to discipline others by controlling self; to keep accounts of what I spend and live within my income, no matter how small; and to bear life's burdens and changes with a cheerful spirit."

'55. "My Holyoke education has helped me for thirty years in performing the duties of a wife and mother, of a Sabbath-school teacher, and helper in all departments of Christian work."

'47. "The value of my school training has been continually rising in my estimation since July, 1847. I was utterly unconscious of the greatness of the debt then, and for years after, but have now seen it for a quarter-century. There I first had a glimpse of what Christian obligations are. There, too, were formed friendships which have endured the strain of more than forty years and are the cherished treasures of my heart to-day."

'70. "No Holyoke graduate can fail to realize more and more the value of the daily discipline of seminary life, in building up strong and Christian womanhood."

'63. "My appreciation of the training there grows with years."

'50. "I prize the religious, moral, and intellectual training received there, and see its value more and more."

'78. "The benefits of Holyoke discipline are appreciated only by degrees. Not until within a few months did I think of the way in which our change of rooms helped to fit us for the discipline of life. We did not remain in one room so long that we could not feel at home in another, and so we are less easily disturbed by the frequent vicissitudes of life. The changes of companions at table and domestic work helped in the same direction."

'79. "The frequent change of roommates teaches one to adapt herself to the peculiarities of those with whom she may journey through life."

'61. "One regulation to which I attach great importance required us to report accounts balanced. The habit then formed I have not relinquished, and I think if all young ladies could see the importance of being able to account accurately for the money they have spent, many a financial crisis would be averted."

'60. "The seminary training forms habits of economy in time, health, and wealth."

'74. "It has helped me to be prompt,—I never was in earlier life; and it has aided me to meet faithfully the disagreeable things that lie in every one's path. If unpleasant work must be done, why should I leave it for some one else, to whom it would be equally distasteful?"

'84. "Life at the seminary makes one more thoughtful. It showed me the importance of method in little things and how to turn odd bits of time to good account."

'79. "In almost every experience, I feel something of its influence to correct or strengthen me."

'48. "My Holyoke training helped me to be a teacher and then a missionary, and taught me to be faithful in little as well as great things—fearing nothing but that I should not know and do all my duty."

'56. "The seminary training did not make of me a woman of distinguished ability in either public or private life, but it helped me as teacher, mother, and member of society, to be faithful in a few things."

'85. "Mount Holyoke education tends to foster thoughtfulness in little things."

'73. "My life has not been tested by great sorrow, but in the way of daily strength for the self-denial involved in serving others, the Holyoke lessons of thoughtfulness have been a blessing."

'39. "Many of its less gifted graduates have done good service for the Master, in the humble walks of life; they have taught the little school, have made the humble home bright; have cared for the sick and trained the young, with an intelligence and refinement far reaching in their results."

'56. "There are better daughters, mothers, and friends, throughout the world to-day, because of the influences of our seminary home, which dignified labor, deepened conviction of truth and duty, and impressed upon us the grand possibilities of life. Eternity alone will disclose how important a factor this institution is and shall be in the evangelization of our world."

'50. "I rejoice that I was one of Miss Lyon's pupils and among the number of those who received her last benediction. My whole life work has been molded by it."

'45. "I consider Holyoke training invaluable for stability of character and steady religious growth. It helped me to train my children, all of whom became Christians early in life and are now active members of the church."

'60. "I am left with five children. It is worth everything to me that I had the discipline of Holyoke to strengthen me for sorrow. The religious life nurtured there keeps me from sinking now that God has laid his hand so heavily upon me."

'61. "The lessons of self-denial inculcated there have made it easier through life to say 'Thy will be done.' I was taught that the highest privilege is service for the Master in the lowliest station."

'52. "The teachings received within those walls strengthened me for the trials and struggles of subsequent years."

'81. "Whatever I am my Holyoke education has helped me to be, and next to mother and home has had the strongest influence."

'69. "In connection with my three years at the seminary, I remember the most distinctly its religious tone. I have always believed that whatever of success I have had in my sixteen years of work as pastor's wife I owe largely to the training I had at Mount Holyoke."

'44. "The systematic habits formed at the seminary were of great value to me in missionary work."

'53. "Holyoke training in punctuality, thoroughness, earnestness of life, the making eternity greater than time, and duty paramount, has given me whatever power for usefulness and efficiency I possess."

'56. "The training tends to make pupils conscientious, faithful, and ready to take up self-denying work; such persons as pastors and their wives like to find—who can be depended on for help."

'60. "I have not entered any profession—nor written any books. Nor have I occupied any large space anywhere, but what I am and what I have been able to accomplish, is due to the seminary. Its influence is inwrought in all I say or do, for it was there I learned to believe in Jesus as my Saviour."

'76. "So suggestive is the course of study, and so thorough the work required, that enthusiastic teachers are the result."

'82. "It is not easy to estimate what Alma Mater has done for me. She has enlarged my sympathies, given me a broader view of the needs of humanity, and taught me how to work. This has been of the greatest value since I went out from the inspiration of numbers to take up my own work in the world. In my education itself I owe a great debt to the seminary, for I could not have had so thorough a course of study elsewhere. I am thankful for the Christian way in which we were taught, because I am constantly meeting theories and questions that tend to skepticism, especially in connection with the natural sciences, and my training there fortified me against their fascination. I shall always be thankful that I was sent there."

'46. "The Holyoke training helps in every department of life. The conscientious truthfulness, the honesty of purpose, the purity of motive inculcated, cannot fail to confer lasting benefit. Christianity and science walk hand in hand, and their harmony is so thoroughly

THE ZOOLOGICAL LABORATORY.
Lyman Williston Hall.

shown that Holyoke pupils can be depended on to stand up for the faith in these skeptical days, and also to give a reason for the hope that is in them. It has been a life-long regret that I could not return to graduate."

'65. "To me the central feature of the seminary training was the use of the Bible. But in the whole course of study—revelation, history, or science—truth was shown to be a unit."

'49. "The zeal that was cultivated there was a zeal according to knowledge. How thoroughly we were grounded in Bible truth."

'62. "The Bible lessons were a spiritual feast."

'80. "The thorough course of Bible study has been very valuable to me in Sabbath-school work."

'77. "My quiet life may seem without fruit, but I am so constantly surrounded by skeptical young men that I have opportunities equal to any mission field, and am thankful now for the drill in the Bible, Evidences, and Butler, though I used to regard much of the time so spent as wasted; but these studies have enabled me to show many a weak spot in the skeptical notions that some young men pretend to believe. When a clergyman told me the other day that one studying for the ministry spoke of me as the first person who caused him to doubt the sufficiency of his skepticism, I felt that it repaid a life-time of obscurity."

'70. "My faith in Christ was greatly strengthened. The teaching at the seminary gives strong support against the infidelity rampant in the world to-day."

'51. "That part of the course of study which bore upon Christian belief has been of untold value to me in meeting doubts from within and skepticism from without. The Bible is the great refuge, yet for these added helps I can never cease to be grateful."

'44. "Whatever of Christian work I have been permitted to do in public or in my family, I attribute to the training of those beloved teachers and the influences of those precious Sabbaths. The putting religious life uppermost, and other things as helps to that, has been invaluable in strengthening me for the Master's work, and for the bereavements, disappointments, and sorrows it has been my lot to bear. Since being so thoroughly established in the 'faith once delivered to the saints,' I have often been able to put aside the whisperings of infidelity and stand firm for the truth, and have thus been able to assist some doubting ones to get a firm foothold on the Rock of Ages. I often wish in these days, when we see so many church members apparently satisfied with the tinsel of the world's amusements, that Miss Lyon was more generally known throughout our land. I have not heard in late years anything definite in regard to the unreserved consecration of her successors to the Master's work; but in its earlier years the seminary left such a distinctive impress upon its pupils, that strangers often said to them, 'Weren't you educated at Holyoke? I thought so, for you have their earnestness.'"

'50. "What a hushed thrill crept through the seminary that March day when it was known that Miss Lyon was dead! It had seemed as if she was essential to the seminary. But she had prepared other faithful laborers to step into the vacant place; and the wheels never stopped. Nor, since that time, has the steady machinery been seriously disturbed. Year after year earnest souls go forth with a training which fits them to do good in the world and meet the vicissitudes of life with a Christian spirit."

'61. "I have laid the circular away a dozen times, half determined not to reply, but when I was in the seminary I did as I was bidden and I cannot disobey now. That is a sacred place to me and will always be, for it was my spiritual birthplace. I had been waiting for a year, very impatiently, to go there and begin the new life. The day we arrived, Miss B——, a new teacher, showed my sister and me to our room. Miss S—— presided at the table and asked a blessing over our late dinner, and her thin, clasped hands, gentle voice, and spiritual face, impressed me deeply. Then Miss Jessup prayed so heartily for the new girls and especially for the homesick ones that night, that I was moved to tears. And when a day or two later she spoke about taking time for prayer, and bade us think how fitting it was that we should begin with our new school life a new life in Christ, my sister and I were completely subdued, and went to our room to do just as she had told us. All through those first weeks Miss B—— lavished on us little attentions that no one expects from a stranger, and that only Christian love suggests. I saw why it was, when, the first Sunday evening, she laid her cheek against mine, while she took my sister's hand and whispered, 'I hope you both belong to the dear Saviour.' It was Miss S——'s privilege to lead me to the light, after weeks of darkness. I never understood how she knew I needed her help, for I could not make my feelings known; but she took me to her room one Sabbath night and without asking a question, told me my trouble and led me to my Saviour. O those Sabbath days, those silent half-hours, and those days of prayer!—how I have longed for their rest and quiet ever since! And in all the years of my Christian work I have been able to think of no wiser, sweeter, or more helpful things to say than were then said to me."

'55. "The one year I spent at the seminary was the most precious of my life. In its quiet halls I found the Saviour, so precious then, and through all the years that have followed."

'86. "I found my All there, and Alma Mater will always be dear to me."

'70. "The memory of those days is very precious to me, for I have never lost the religious interest then awakened."

'68. "It taught us to place religion first in everything."

'64. "I learned that one need not be known to fame in order to be a power for good. Others may tell of beloved teachers and of the influence of seminary training; I want to speak of the effect left on me by observing the unswerving integrity of that most obliging man of all work,—faithful Cornelius,—who never lost his temper though wanted in all parts of the building at the same time, and who cared for all the interests of the seminary as if they were his own. I can never forget the lesson taught by the delicacy of his Christian sympathy one sorrowful day when he turned aside in the domestic hall to tell me how sorry he was to hear that my soldier brother would never come back from the war. And when (in 1883) I heard that after nearly thirty years at the seminary he had retired to his little farm, I said, 'It will be hard to find another to fill the place of Cornelius Rutherford,—a man worthy of both his names.'"

'57. "That revival year I was strongly impressed with the consistent Christian lives of the older pupils, about their domestic work, and everywhere."

'59. "It was the general standard of Christian uprightness that left the most pronounced impression on my mind."

'47. "I regard the decision to enter Mount Holyoke Seminary as the most important one of my life. I had previously made public profession of my faith in Christ. But I did not know the full meaning of that act till I saw the consecrated lives of Miss Lyon and her associates. The daily study of God's Word and its wonderful application to the daily life was set before me as never before. I was almost in a new world. I had so longed for the higher education there afforded and for the full development of all my powers, for which everything at the seminary was admirably fitted, that at once I accepted most gratefully the entire system. The family work was a pleasure, for the young ladies entered into familiar and yet congenial intercourse, while it continually enabled us to bear one another's burdens in many appreciative ways. The self-reporting system led every pupil to watch over her own actions, and always to know whether she was in the right place and making the best use of her time. This watchfulness and care, the regular exercises of school and family arrangements, necessitating 'schedules of time' made out and lived up to, all formed habits of system whose value could not be over-estimated. But the religious life of the school as led by Miss Lyon and her teachers made the place a perennial fountain of blessing. Very few, after spending three years in such training, could go home to settle down into listless inactivity or aimless lives. Christian women in every department of life, with the mental and moral culture gained at South Hadley, are blessing our homes and schools, the church and the world."

'51. "Holyoke training is remarkably adapted to strengthen character, and lead to useful and practical after life. I remember with gratitude the tender religious influences which almost impelled us to consecration to God. Nor can I forget the subsequent influences, which encouraged us to lead in prayer from the very first, and to try to win our friends to Christ. I am confident that these early lessons have followed many into later years, making Christian service sweet, and self-denials for Christ a delight."

'40. "I cannot be thankful enough for what my two years of Holyoke training did for me. It helped me to be more self-reliant, to make sacrifices willingly for the good of others,—in a word to feel that I must live to do some good in the world. It has helped me to train my children for usefulness. Miss Lyon's influence is going on like the waves of the sea and eternity alone will reveal the good she has done.

"Do not think me egotistical if I give an incident. Two years ago I went as a delegate to a temperance convention in P——, forty miles distant, and was urged to take the place of an absent speaker. I felt that I could not, but a lady near me said, 'Go on, you need not make a long speech.' Breathing a silent prayer for help, before I reached the platform the words of Miss

Lyon came to my mind, 'Be willing to go where no one else will go and do what no one else is willing to do.' I gave that as a reason for coming forward before so large an audience with such short notice, and words seemed given to me as I needed them. A year later I attended another temperance convention and was asked to address the meeting. When I declined, the president said with great earnestness, 'If you only knew how much good those words you quoted at P—— had done me you would not refuse.' She said, 'I have been called to labor among rough, drinking men, and my heart often shrank from the task, but those words of Miss Lyon continually came to me with such force I dared not refuse, and my labors have not been in vain.'"

'69. "It was my seminary life that freed my tongue, and made me feel that I could say a word for the Master when occasion offered. I used to think religion was something that one could not talk about; but if I have been led to say anything since, that has helped any in the journey heavenward, it was gained at the seminary; and you may put that in any book you like, if you will only put it in a better shape. I often feel as if I were an idler in the Master's vineyard—but I do love to speak a word for him, though I have not always the courage or the skill to make the opportunity."

'58. "Representing, as we did, all the states and territories of the Union—not to mention foreign countries—we gained from each other an important part of our education; and though New England principles prevailed, the influence of the seminary was neither narrow nor provincial."

'72. "I owe much to the unsectarian piety of the seminary. Brought up a strict Presbyterian and having naturally strong prejudices, my sympathies were greatly broadened there, and I have never been able to consider sects essential since our little school world maintained a thorough-going Christianity on a non-denominational basis."

'57. "I look back on my days at the seminary as blessed days, fitting me for a life of care and struggle by the development of faith in Jesus as my Saviour. Waking and sleeping I live over my school days. The firm religious impressions, the thorough instruction, and the practical ideas of order, then received, have been of incalculable value to me these thirty years."

'40. "For myself, as doubtless for many of the younger pupils, Holyoke training laid the foundation for subsequent education, showing the value of all knowledge, and stimulating to effort for its acquisition."

'80. "Its whole education has made me feel unsatisfied with present attainments and desirous of pressing on in the life work open to every Christian woman."

'41. "A Holyoke education leaves a marked impress. Aside from its intellectual training it creates a high resolve 'to be' through life a blessing in the world; 'to do' for others whatever may be given to do; and 'to bear' silently and heroically, whatever toil, privation, or suffering this may involve."

'38. "I was at the seminary only the spring term of the first year, but I would not have been denied the privileges of that one term for any money. The stamp on every one I have met from that home seems to be, 'There is a great work to be done, and I will do what I can'; it is the attitude of the good soldier—'on duty, and ready for further orders.'"

'41. "Holyoke influence has tended to develop in me independence of character and a greater sense of individual responsibility, so that for the most part the question, what is right, or duty—not what will people think—is paramount; and I have endeavored to stamp the same on our children and I think it has saved them from many a snare."

'48. "The strongest impression left upon Holyoke graduates is a deep sense of personal responsibility, and a greater readiness to give themselves to Christian work."

'59. "The strongest effect was the recognition of personal responsibility to do or suffer, whatever was to be done or suffered. I believe Holyoke graduates, as a rule, are strong in their devotion to duty, and uncompromising toward evil."

'61. "For myself, Holyoke influence had much to do in making seeming sacrifice become positive pleasure. The impress upon its pupils is that of personal responsibility, the duty of fitting one's self to be a sharer in the world's work."

'63. "Holyoke education, which I regard emphatically as Christian education, helps many to do for the poor and ignorant so far as in them lies, that which has been done for them, striving to bear one another's burdens and so fulfill the law of Christ."

'69. "My three years in the school gave a new direction to my life, and a desire to work. The distinctive imprint of the school is the desire to accomplish something worthy of a Christian woman; to help those who need help."

'75. "A very distinct impression was made upon me by the ambition of so many to fit themselves to work for God in all the walks of life."

'49. "Years ago I met a lady who said, 'I do not like Mount Holyoke graduates, for they always feel that they have a mission, and they cannot rest till they have accomplished it.'"

'58. "Holyoke training leaves upon its pupils the stamp of fixed purpose, and earnest Christian endeavor. I have been largely indebted to it in efforts to inculcate on my pupils a desire to act at all times from Christian principle."

'53. "I am filled with increasing gratitude to the seminary for the grand ideal of life which it gave me. The lesson, from Alpha to Omega, of living for God, of consecrating all to him, is prominent in every recollection of its teachings."

'45. "I think the distinctive impress is a love of thorough work and a desire to do all things heartily, as unto the Lord."

'77. "I learned that to walk in Christ's way is not only safest, but the only way to satisfy an enlightened conscience, and ordinary common sense. That year stands out from the other years of my life like a cameo from its setting."

'64. "The result is unswerving adherence to duty, after careful and conscientious study to know it."

'49. "Holyoke training leaves its mark upon its pupils. It is sometimes said, 'Character is its specialty.'"

'54. "The seminary produces a certain equipoise of the faculties, the outgrowth of the system for which it is justly noted."

'56. "The regular home duties to be performed as faithfully as George Herbert's 'sweeping of a room'; the instructions of those faithful teachers; the quiet season of devotion at the close of the evening meal, coming in like a benediction after the toils of the day; the little recess meeting, where souls first learned to pray with and for each other; the half-hour alone; and, above all, the spiritual atmosphere that brooded over all the place, have left an impression on me that nothing can efface."

'77. "I sometimes think the greatest power for good lay in the silent half-hour. Ever since, I have had a deeper sense of the necessity of spending some time each day alone with God."

'74. "If there is one thing of more practical value than any other, it is the habit there formed of regularly setting apart a stated time each day to be alone with God and his word. This is most restful, and of the greatest help, not only in preparation for whatever one may be called to meet through the day, but in promoting steady spiritual growth."

'45. "Perhaps there is nothing that I look back on with more gratitude than the provision made for the silent half-hour; and I have often wished in after life that there was a bell that would shut out all interruption for half an hour each day."

'75. "I can never speak highly enough of the morning and evening 'half-hours.' To me they were the best parts of the day."

'57. "The influence of the half-hour of prayer has never departed from me. The habit then formed has been of untold value to me, and indirectly to my kindred."

'59. "The influence of the silent half-hour, the section prayer meeting, and the instruction given at family prayers, is beyond all estimation. We do not know how deep and thorough that education was, until it has been tested in after years."

'45. "I thank God that I went to Mount Holyoke Seminary, where the training of Christian parents was continued by Christian teachers."

'45. "I am more thankful to my Heavenly Father for allowing me to be one of Holyoke's daughters than for any other temporal blessing. It has been a life-long disappointment that I could not stay to graduate."

'47. "I shall ever be thankful for the missionary influence which pervaded life at the seminary. The atmosphere seemed full of it, and the lights and shadows in the lives of missionary ladies in Persia seemed to cloud or brighten our lives as well. I well remember Miss Lyon's saying one day, 'Yes, young ladies, as the millennium draws near, Christians will cast their luxuries into the Lord's treasury, gladly curtail their comforts, and even look closely upon the necessaries of life, that they may give the more.' I have tender memories of the Sunday morning prayer meetings when we recited to each other the names and location of the various missionaries of the American Board and sought to hold them up before God in prayer. Others will ever remember the day appointed for special prayer for foreign missions when the 'beloved Persis' (Thurston) led our section meeting and told us some of the experiences of her father in Hawaii. I, for one, received an impulse which intensified my long cherished wish to be a foreign missionary. You will, perhaps, remember Miss Lyon's characteristic statement of qualifications essential for a missionary,—'first, of course, piety; next, a sound constitution; and then, a merry heart.'"

'47. "Our first missionary meeting was made especially attractive that we might become interested in a subject of which most of us knew but little. Our 'beloved Persis' dressed as a Hawaiian, sought to interest us in the land of her birth. It made a deep impression on me, and from that day I loved the missionary work. In the church of which I was a member twenty-five years ago, I was the only one besides our pastor's wife, who had any interest in missions. This church is now a center of active service, and if I was of any help in creating a missionary spirit, it was all due to the teaching of Miss Lyon. She never knew that she had influenced me for good, for I gave no evidence of it while with her. It will be one of her 'sweet surprises' hereafter to learn that she had done so much for me. To the systematic training of Holyoke I am infinitely indebted in all the walks of life, and I am glad to offer my testimony to its worth."

'42. "Miss Lyon's earnest wishes have been fulfilled in successful workers in many lands, who attribute much of their usefulness to the principles inculcated at Mount Holyoke Seminary."

'67. "I am very thankful to have had my interest in mission work stimulated while at Holyoke by meeting so many who had been in the field. Especially was it a rare privilege to be at the seminary when Miss Fiske was there."

'59. "I could not have continued in our missionary work, without the oneness of purpose obtained at Mount Holyoke."

'71. "My school life has earnestly interested me in the great moral questions of the day, and made me realize that upon the Christian women of the land is laid a responsibility which God alone can enable them to carry."

'44. "The missionary spirit burning in the heart of Mary Lyon has kindled an answering zeal in the hearts of thousands. I trace my own long interest in missions directly to the influence of her teachings."

'49. "How could any one be a pupil of hers without feeling that she had a work of her own to do for the conversion of the world? In reviewing the history of the 'Woman's Boards' in our country, it will be found that most of the original leaders were pupils of Mary Lyon, and many of their successors were taught by her pupils. I believe that her influence gave the first impetus to this great movement."

'57. "I have found Holyoke alumnæ members of the executive committee of more than one Woman's Board, and their most efficient helpers. The knowledge of Miss Start's self-sacrificing labors as city missionary made me less selfish, for she was far less fitted to endure hardness than I."

'83. "My Holyoke training stimulated an interest in church and mission work."

'85. "If there is one thing more than another that I gained at Mount Holyoke, it is a deep interest in and love for mission work."

'48. "Thirty-five years have passed and my daughter has graduated and is now with us on missionary ground."

'51. "There I sent my only daughter who left the seminary a quarter of a century after her mother, and I have no doubt she will continue to show the influence of the training received there."

'59. "The best proof of my sincere devotion to Holyoke is to be found in the fact that my only daughter is a senior there now. I have been more than satisfied with the results of her connection with the seminary."

'53. "My confidence in Mount Holyoke is abiding. Perhaps I have given the best proof of this in sending two daughters there, and I hope to send a third."

'46. "Perhaps no better testimony can be given to my interest in the seminary, than the fact that two daughters have already completed the course, and the remaining one is soon to graduate."

'57. "My Holyoke education has been the inspiration of my life, and if I had twelve daughters, I should wish them all educated there, as my only one has been. I was glad to hear that there were thirty daughters of alumnæ in the seminary in 1886, and that the number this year is larger still."

'53. "Our daughters love the seminary next to their own home."

'62. "My great sorrow is that I cannot send my boys there to be educated."

'61. "I cannot refrain from expressing my gratitude to Holyoke that she placed her tuition so low."

'57. "With me an education in any other first-class institution would have been impossible on account of the expense. Most of my bills were paid from my own earnings."

'61. "I wonder if Miss Shattuck remembers a 'new scholar' asking her how long she had been in the seminary, and to her reply, 'ten years,' exclaiming, 'What, not through yet?' Yes! it is because of such life devotion and energy as she has shown in her department, that our Alma Mater stands where she does to-day; and we bless God that she and other dear teachers are 'not through yet!'"

'55. "With alacrity we obeyed Miss Jessup's request 'to step to the basement' for an errand, though we were in the fourth story, or *vice versa*. Dear Miss Jessup, how we would still step for her if we might loose the chains that have bound her for so many years!"

'58. "I do not know that I should have my present Sunday-school class—six boys about the age of fourteen, who had driven away three gentlemen in succession—had it not been for Miss Jessup's oft repeated 'Do what no one else will do.'"

'47. "The ideal of life, which I have cherished ever since, was largely furnished me by Mary Lyon. Through all the events of my life, 'What would Miss Lyon say?'—'What would Miss Lyon do?' have been, under Christ, the questions that have guided me. I think I have observed the same influence in many other graduates. I specially rejoice to observe also that some of the pupils of later years, who never knew Miss Lyon, are under the control of an ideal which the seminary has succeeded in perpetuating from her, so that she 'being dead yet speaketh.' May the seminary never fail to preserve that sacred inheritance."

'52. "I was only sixteen and full of life and fun, but there I found Christ. There was always an influence in the school that was drawing me slowly but surely to him. I believe it is there still, a constant answer to all the fervent prayers that from Miss Lyon first, and from so many ever since, have been going up from within its walls."

'55. "We can never cease to be grateful that our Alma Mater is still pleading for us, and when sorrow comes we feel strengthened to bear it with Christian fortitude."

'70. "The assurance that I am remembered among 'all those who have been of our household in times past' is a great comfort to me."

'39. "Prayers for its prosperity and advocacy of its distinctive features have borne witness, these many years, to what Mount Holyoke did for me."

'40. "My most earnest prayers are for the future extended usefulness of dear old Holyoke."

'63. "The seminary has a peculiar attraction to all former pupils. A visit to her sacred walls is like returning to the paternal roof and asking for one more blessing. A sacred tie exists between those educated there, which binds our hearts together however far we may be sundered."

'61. "I remember with gratitude the delightful family feeling existing between teachers and pupils."

LECTURE ROOM.
Lyman Williston Hall.

'58. "I love to recall the courteous 'good morning' with which Miss Chapin was sure to greet us as we met in the halls, never failing to remember our names. This was one of the many ways by which she made us feel her personal interest in each of us."

'63. "If 'the reward of work well done is, not rest, but more work,' then truly did Miss Chapin prove her service acceptable to the Master."

'73. "Mount Holyoke Seminary was to me more than a school; it was a Christian home, filled with all gentle and sanctifying influences."

'86. "I would not exchange my four years there for a course in any other school in the land."

'39. "None of the younger pupils can love our Alma Mater as do we who sat at Miss Lyon's feet, and entered into her endeavors. The progress of the seminary means more to us, who see in it a fulfillment of the grand ideal of those early days."

'46. "It is cheering to note the progress from year to year."

'65. "I rejoice in its present prosperity and that it is so well keeping pace with the progress of the age."

'67. "Thrice glad are we for all the improvement wrought in so many ways."

'61. "We have noted with thanksgiving that our Alma Mater has been kept steadfast in the service of Him to whom it was first dedicated. May the gracious blessing of the Master ever rest on all who gather in the dear home."

'66. "All my home loves and cares do not drive out my love for the seminary and the friends there. I am thankful for the years spent within those walls, and if God should ever put it in my power, one of my greatest pleasures would be to help build up the seminary to the ideal position which all her children desire for her. My heart will be with you at the jubilee, but I cannot be there in person."

'47. "I wish to add my voice to the many that will unite this jubilee year in a chorus of gratitude."

'48. "Ever since the first note of preparation for the jubilee was struck, I have proposed to be there. If I cannot enjoy the coveted privilege of being within the walls of my beloved Alma Mater, then give me a cot close by, that I may look upon her fair structure the last thing at night and the first in the morning. For it will be the last visit I shall ever make to that dear spot."

'42. "How I would love to be present at the semi-centennial! I shall be with you in spirit, but I never expect to see the dear old home again. May there be a happy reunion, and the blessing of the Lord rest upon it!"

'43. "I am nearly blind—but I want a copy of the 'History' and will get some one to read it to me."

'40. "Probably I shall not live to see the jubilee or read the book—but I wish to subscribe for a copy for my son."

'40. "May you have a delightful time at the jubilee! There will be few to represent the first pupils who gathered in the dear old home. Some will send loving thoughts from afar. I wonder if Miss Lyon and Deacon Porter will not be permitted to look on. It may be there will be 'a great cloud of witnesses.' You remember Miss Lyon said in her last illness, 'I should love to be permitted to come back and watch over the seminary,' and then added—and we all echo the words —'but God will take care of it.'"

Let this chapter of testimony close with the words of Dr. Kirk at the reunion on the twenty-fifth anniversary:—

"This unusual gathering of teachers and alumnæ shows how much this school is beloved of God's children. . . . It is just what Mary Lyon intended it to be; what she constantly prayed it might be: stronger than when she left it; independent of her or any other individual human being; living in the favor of God and the confidence and sympathy of the churches. There cannot be a question that it is already a vast power in our nation, both as a part of the great educational machinery of our country, and as founded on lofty Christian principles. The question that awakens our solicitude is, Will its guardians prove faithful to their trust? It is no common piety that can keep it where its founder left it. A declension of piety in the churches would come stealthily up here like the miasma of death. If the church ceases to pray for it, the blessing of God will be proportionally diminished. This seminary is a sacred trust from the Lord to his churches. If the trustees should come to regard it in a merely secular light, their influence will be hurtful rather than beneficial. If the teachers should come down from the high places of prayer, of close walking with God; if worldly ambition and self-seeking should gain possession of their hearts,—it would so far fail of its original design, and the most sacred of trusts be so far betrayed.

"Fellow trustees! We are guardians of an institution dear to many now in heaven; dear to Him that sits upon its throne. Let us to-day take a new view of our trust and watch ourselves lest we impair some element of its strength. Teachers within those walls! You are laboring for Christ and eternity. The instant you come down to the common ground of self-seeking you part company with Mary Lyon and you betray her dear seminary. May her Saviour keep you as he kept her! Daughters of Holyoke! Refresh yourselves with the precious memories of the past. It does a generous soul good to revive its sense of obligation. Remember the goodness of God and praise him. Remember the true hearted ones that so nobly bore the burden and the reproach of this enterprise in its infancy. Bless God for their noble work; greet each other here once more; and then go forth to carry out even more fully the sacred principles you were taught here.

"May our blessed Lord make the coming years witness even richer blessings on the school than all our eyes have yet witnessed! To him we commend its sacred interests!"

CHAPTER XX.

ALUMNÆ AT WORK.

THE term alumnæ as used in this volume includes all former pupils,—graduates or not. Some hints of their varied work are given in the preceding chapters. The records of the Memorandum Society show what positions are filled by the members of that body. The number reported in its catalogue at the end of forty years—ten years ago—was 2,341, and included 737 besides the 1,604 graduates. This was less than half the whole number—about 4,750—that had then been connected with the seminary.

Of these 2,341, 1,690 had taught since leaving the seminary; 77 of them twenty or more years; 260 between ten and twenty years; and 470 between five and ten years; making 807 that had taught five or more years; 21 were physicians; 1,391 were married; 141, perhaps more, were or had been foreign missionaries. The number of city and home missionaries, known to be large, was not definitely ascertained.

The secretary is endeavoring to obtain similar information for the whole half century, extending her inquiries to include other work, philanthropic, educational, and literary; and not only about those belonging to the society but respecting all that have ever been members of the seminary; but the work is too great to be finished in season for these pages. Till that record is complete, it will be impossible to give the names or approximate number of home missionaries, or even of the teachers laboring among freedmen, Mormons, Indians, or Chinese in our own land. Nor can the full number of foreign missionaries be given.

ALUMNÆ AT WORK.

The subjoined list of those under foreign boards is inserted in the hope that readers who notice omissions or errors will kindly report them to the secretary, Mrs. C. B. Pease, Somers, Connecticut, who will be glad to receive information also about unreported alumnæ in any part of the world.

ALUMNÆ WHO ARE OR HAVE BEEN IN THE SERVICE OF FOREIGN MISSIONARY BOARDS.

The figures at the left indicate, with names of graduates, the year of graduation; with undergraduates, the year of leaving the seminary: d. with date denotes death.

It is not attempted here to specify changes, nor to indicate the present field of labor.

'54. Augusta E. (Abbott) Dean, Ahmednagar, India.
'44. Abby (Allen) Fairbank, Ahmednagar, India; d. 1852.
'50. Anna C. (Allen) Douglass, Madras, India.
'68. Martha A. Anderson, Ahmednagar, India.
'79. Fanny P. (Andrews) Shepard, Aintab, Turkey.
'41. Harriet (Arms) Sylvester, Choctaw Nation; d. 1868.
'48. Maria P. Arms, Choctaw Nation.
'39. Mary (Avery) Loughridge, Creek Nation; d. 1850.
'48. Lydia H. (Babbitt) Dodd, Marsovan, Turkey.
'38. Charlotte (Bailey) Grout, Natal, South Africa.
'68. Louisa M. (Bailey) Whitney, Ebon, Ralik Islands.
'68. Isabella C. (Baker) Stocking, Oroomiah, Persia.
'39. Elizabeth K. (Baldwin) Whittlesey, Hawaiian Islands.
'66. Anna M. (Ballantine) Park, Bombay, India.
'57. Elizabeth D. (Ballantine) Harding, Sholapur, India.
'55. Mary (Ballantine) Fairbank, Ahmednagar, India; d. 1878.
'56. Cornelia C. (Barrows) Bartlett, Cesarea, Turkey.
'62. Martha J. Barrows, Kobe, Japan.
'45. Lydia (Bates) Grout, Natal, South Africa.
'59. Aura J. Beach, Oroomiah, Persia; d. 1884.
'59. Charlotte (Birge) Chamberlain, Arcot Mission, India.
'81. Emily R. Bissell, Ahmednagar, India.
'79. Julia Bissell, Ahmednagar, India.
'81. Ellen M. Blakely, Aintab, Turkey
'39. Emma L. (Bliss) Van Lennep, Smyrna, Turkey; d. 1840.
'63. Flavia S. (Bliss) Garner, Marsovan; Sivas, Turkey.
'47. Georgiana M. (Bliss) McQueen, Corisco, West Africa.
'56. Caroline (Boynton) Kingsbury, North American Indians; d. 1873.
'54. Celestia (Bradford) Carleton, Ambala, North India; d. 1882.
'83. Ella T. (Bray) Graham, Aintab, Turkey.
'52. Susan A. (Brookings) Wheeler, Harpoot, Turkey.
'55. Mary L. (Browning) Herron, Landour, North India; d. 1863.
'69. Abbie L. (Burgess) Hume, Ahmednagar, India; d. 1881.
'66. Mary A. Burnett, Peking, China.
'69. Theresa M. Campbell, Alexandria, Egypt.

'70. Mary L. (Carpenter) Howland, Mandapasalai, India; d. 1887.
'52. Maria J. (Chamberlain) Forbes, Hawaiian Islands.
'51. Sarah A. (Chamberlain) Scudder, Arcot Mission, India; d. 1870.
'41. Malvina J. (Chapin) Rowell, Hawaiian Islands.
'42. Martha R. (Chapin) Hazen, Ahmednagar, India; d. 1884.
'50. Annie S. (Chase) Willey, Cherokee Nation; d. 1862.
'52. Anna M. (Child) White, Madura, India; d. 1877.
'53. Ann E. (Clark) Gulick, Okayama, Japan.
'64. Ursula E. (Clarke) Marsh, Philippopolis, Turkey.
'72. Virginia (Clarkson) Cady, Kioto, Japan.
'69. Elizabeth (Cobleigh) Cole, Bitlis, Turkey.
'46. Hannah Maria (Condit) Eddy, Beirut, Syria.
'42. Marcia Colton, Choctaw Nation.
'51. Mary J. (Crofut) Morse, Bangkok, Siam.
'72. Ulee P. (Cross) Crumb, Toungoo, British Burma.
'46. Mary M. (Curtis) Seymour, Choctaw Nation; d. 1859.
'74. Anna Y. Davis, Kobe, Japan.
'57. Elizabeth A. (Davis) Greene, Constantinople, Turkey.
'47. Eunice B. (Day) Bliss, Constantinople, Turkey.
'48. Eliza M. (Dewey) Pierce, North American Indians.
'54. Elizabeth (Diament) Canaday, Choctaw Nation.
'54. Mary (Diament) Ramsay, Seminole Indians.
'54. Naomi Diament, Creek Nation; Peking, China.
'42. Caroline E. (Dickinson) Bissell, Choctaw Nation; d. 1876.
'66. Sarah E. (Dyer) Pierson, Pao ting fu, North China; d. 1882.
'61. Charlotte E. Ely, Bitlis, Turkey.
'61. Mary A. C. Ely, Bitlis, Turkey.
'65. Olive J. (Emerson) Morrow, Tavoy, British Burma.
'77. Katie Fairbank, Ahmednagar, India.
'77. Caroline P. (Farnsworth) Fowle, Cesarea, Turkey.
'82. Anna Felician, Marsovan, Turkey.
'56. Joanna (Fisher) White, Oorfa, Turkey.
'42. Fidelia Fiske, Oroomiah, Persia; d. 1864.
'44. Nancy A. (Foote) Webb, Madura, India.
'65. Mary J. (Forbes) Greene, Kioto, Japan.
'78. Sarah A. Ford, Sidon, Syria.
'52. Eliza J. (Foster) Scott, Landour, North India.
'55. Sarah J. (Foster) Rhea, Oroomiah, Persia.
'63. Nancy D. (Francis) Adams, Aintab, Turkey.
'57. Ann Eliza Fritcher, Marsovan, Turkey.
'68. H. Juliette Gilson, Natal, South Africa.
'67. Alice (Gordon) Gulick, San Sebastian, Spain.
'66. Mary E. Gouldy, Osaka, Japan.
'86. Anna D. Graham, Aintab, Turkey.
'56. R. Oriana (Grout) Ireland, Natal, South Africa.
'40. Lois W. Hall, Cherokee Nation; d. 1861.
'57. Margaret E. (Hallock) Byington, Constantinople, Turkey.
'46. Eliza (Harding) Walker, Diarbekir, Turkey.

'82. Alice B. (Harris) Smyth, Foochow, China.
'76. Emily S. Hartwell, Foochow, China.
'58. Lucy E. (Hawley) Ing, Kiu Kiang, China; d. 1881.
'75. Francis A. (Hazen) Gates, Sholapur, India.
'41. Sophia D. (Hazen) Stoddard, Oroomiah, Persia.
'55. Louisa (Healy) Pixley, Natal, South Africa.
'76. Anna B. Herron, Landour, North India.
'78. Mary A. Holbrook, M. D., Tung-cho, China.
'67. Mary G. Hollister, Hadjin, Turkey.
'77. Carrie E. (Hoover) Bushell, Rangoon, British Burma.
'49. Angelina (Hosmer) Carr, Choctaw Nation; d. 1864.
'70. Susan R. Howland, Oodooville, Ceylon.
'64. Charlotte E. (Hubbard) (Penfield) Devins, Madura Mission, India.
'56. Emma M. (Hughes) Roberts, Shanghai, China.
'65. Sarah J. Hume, Ahmednagar, India.
'46. Harriet (Johnson) Loughridge, Creek Nation.
'75. Mary A. (Kelley) Leavitt, Osaka, Japan.
'76. Leila (Kendall) Browne, Harpoot, Turkey.
'68. A. D. H. Kelsey, M. D., Tung-cho, China; Hierosaki, Japan.
'48. Abby L. (Kingsbury) Kerr, Canton, China; d. 1855.
'48. Celestia A. (Kirk) (Maynard) Edson, Salonica, Turkey.
'69. Anna R. (Kuhn) Weaver, Bogota, South America.
'48. Abby T. (Linsley) Wilder, Natal, South Africa.
'57. Mary E. (Linsley) Goodale, Adana, Turkey.
'78. Sarah E. (Lyman) Holbrook, Natal, South Africa.
'40. Lucy T. (Lyon) Lord, Ningpo, China; d. 1853.
'78. Lillian E. (Mateer) Walker, North China.
'79. Helen E. Melvin, Constantinople, Turkey.
'45. Sarah P. (Merrill) Bacheler, Midnapore, India.
'38. Abigail (Moore) Burgess, Ahmednagar, India; d. 1853.
'59. Esther E. (Munsell) Thompson, Oroomiah, Persia.
'50. Rose H. (Murphy) Edwards, Choctaw Nation; d. 1881.
'56. Laura B. (Nichols) Bridgman, Natal, South Africa.
'66. Roseltha A. Norcross, Eski Zagra, Turkey; d. 1870.
'59. Zoe A. (Noyes) Locke, Philippopolis, Turkey.
'84. Mrs. Mary A. Oldham, Singapore, East Indies.
'48. Eliza P. Otis, North American Indians.
'70. Mary L. Page, Smyrna, Turkey.
'61. Olive L. (Parmelee) Andrus, Mardin, Turkey.
'73. Alice C. (Parsons) Ballantine, Ahmednagar, India; d. 1878.
'63. Ellen C. Parsons, Constantinople, Turkey.
'68. Lavinia (Peabody) Pearce, Madras, India.
'40. Abigail Peck, Tuscarora Indians.
'66. Jane S. (Peet) Macgowan, Amoy, China.
'83. Ellen L. (Peet) Hubbard, Foochow, China.
'50. Elizabeth W. (Penny) Wood, Ahmednagar, India.
'83. Fidelia Phelps, Natal, South Africa.
'68. R. Ellen (Pierce) Pitkin, Bogota, South America.

'47. Alzina V. (Pixley) Rood, Natal, South Africa.
'48. H. Louisa (Plimpton) (Peet) Hartwell, Foochow, China.
'56. Emily (Pomeroy) Bissell, Austria.
'57. Clara C. (Pond) Williams, Mardin; Constantinople, Turkey.
'68. Harriet G. Powers, Erzroom, Turkey.
'70. Martha E. Price, Natal, South Africa.
'39. Susan (Reed) Howland, Oodooville, Ceylon.
'61. Feronia (Rice) Carpenter, Labrador.
'46. Mary S. Rice, Oroomiah, Persia.
'40. Prudence (Richardson) Walker, Gaboon Mission, West Africa; d. 1842.
'39. Martha (Sawyer) Burnell, Melur, India; d. 1883.
'79. Hettie E. Scott, Landour, North India.
'69. Sarah E. (Sears) Smith, Marsovan, Turkey.
'40. Zeviah L. (Shumway) Walker, Gaboon Mission, West Africa; d. 1848.
'44. Eliza J. (Smith) Wilder, Kolapur, India.
'48. Elizabeth A. (Smith) Noyes, Pulney Hills, India; d. 1880.
'42. Abby M. (Stearns) Cummings, Foochow, China.
'69. Flora P. (Stearns) Bowen, Manissa, Turkey.
'49. Lucy E. (Stearns) Hartwell, Foochow, China; d. 1883.
'45. Persis G. (Thurston) Taylor, Hawaiian Islands.
'59. Martha W. (Tinker) Raynolds, Van, Turkey.
'45. Susan L. (Tolman) Mills, Batticotta, Ceylon.
'55. Sarah L. (Utley) Woodin, Foochow, China.
'70. Helen M. Van Doren, Amoy, China.
'73. Mary L. (Van Meter) Kelley, Maulmain, British Burma.
'61. Mary L. (Wadsworth) Bassian, M. D., Constantinople, Turkey.
'57. Elvira M. (Wait) Dodge, Mendi Mission, West Africa.
'62. Louise (Walker) Gaines, Kioto, Japan.
'60. Fannie E. Washburn, Marsovan, Turkey.
'59. Cora A. (Welch) (Tomson) Millingen, Constantinople, Turkey.
'55. Caroline R. (Wheeler) Allen, Harpoot, Turkey.
'70. Emily C. Wheeler, Harpoot, Turkey.
'40. Maria K. (Whitney) Pogue, Hawaiian Islands.
'72. Cornelia P. (Williams) Chambers, Erzroom, Turkey.
'72. Clara G. Williamson, Landour, Northern India.
'55. Eliza D. (Winter) Morse, Eski Zagra, Turkey.
'46. Celia S. (Wright) Strong, Cherokee Nation; d. 1850.
'64. Lucy (Wright) Mitchell, Oroomiah, Persia.

The four whose names follow engaged in mission work in the places named, but may not have had a formal appointment.

'50. Paulina (Avery) Woodford, Cherokee Nation; d. 1858.
'53. Martha A. J. Chamberlain, Honolulu, H. I.
'50. Ellen R. (Whitmore) Goodale, Cherokee Nation; d. 1861.
'50. Sarah (Worcester) Hitchcock, Cherokee Nation; d. 1857.

ALUMNÆ AT WORK.

Since 1853 the work in the Hawaiian Islands, and since 1860 that among North American Indians, has been continued on a home missionary basis; but it should be borne in mind that parts of our own land were once more difficult to reach than the heart of Africa to-day.

If to the preceding names be added those of the twenty-four who are or have been teaching in South Africa, the whole number given will be 197. Although these teachers are not connected with any board, they were called as missionaries and responded in the same spirit. Their names are as follows:—

'74. Minnie F. Bailey, Wellington.
'81. L. Jennie Baker, Worcester.
'62. Anna E. Bliss, Wellington.
'69. Theresa Campbell, Riversdale.
'63. Susan M. Clary, Pretoria.
'76. Mary E. Cummings, Wellington.
'68. Mary F. Farnham, Stellenbosch.
'56. Abbie P. Ferguson, Wellington.
'85. Margaret E. Ferguson, Bloemfontaine, Orange Free State.
'68. H. Juliette Gilson, Stellenbosch.
'74. Carrie E. Ingraham, Stellenbosch.
'75. Mary E. Landfear, Wellington.
'76. Sarah J. Lester, Stellenbosch.
'79. Evelyn Metcalf, Stellenbosch.
'75. Martha C. Newton, Wellington.
'81. Elizabeth F. Post, Graaf Reinet.
'79. Mary O. Preston, Wellington.
'78. Addie L. Reed, Graaf Reinet.
'73. M. Theodosia Ruggles, Pretoria.
'72. Virginia G. (Sloan) Peast, Stellenbosch.
'73. Ellen A. Smith, Worcester.
'57. Angeline L. Steele, Stellenbosch.
'60. Sarah A. Thayer, Graaf Reinet.
'67. Annie M. Wells, Wellington.

The reflex influence of these widely scattered daughters cannot be estimated. They do not forget the seminary. That their love includes its material as well as its spiritual interests is shown by their contributions to its cabinets. Of these, mention is thus made by Miss Lydia W. Shattuck, as quoted by Dr. Laurie in "*The Ely Volume; or The Contributions of our For-*

eign Missions to Science and Human Well-Being."
Page 176:—

"Our gifts from missionaries have been so numerous and have extended through so many years, that it is almost impossible to give a full account of them. When I began the botanical collection here, I found hundreds of plants in the bundles just as they were sent. I have had them put up carefully in large tin boxes, but I do not know in every case who sent them. From China, Ceylon, Persia, Palestine, Turkey, Spain, Africa, Labrador, and some of our North American Indian missions, many valuable collections of plants, woods, and seeds have come to us, and beautiful collections of algæ and ferns have been sent from numerous localities. In the department of zoölogy, we have from Africa, birds, serpents, fishes, shells, eggs, insects, and horns and skins of quadrupeds; from India, shells and birds; from the Marshall and Sandwich Islands, shells and corals; and the same from Burma, China, and Japan. Rev. Mr. Bruce and Rev. Dr. Fairbank have sent hundreds of specimens, both in zoölogy and botany. Minerals have been received from India, Sandwich Islands, Spain, Persia, and Japan.

"It would leave an immense gap in all our cabinets to take away our missionary treasures. The incidental work done by our devoted missionaries for the advancement of human knowledge would compare favorably with all that governments have done who have made that the sole object of national exploring expeditions."

In this connection should be noticed the work of Miss Shattuck herself, the worthy successor of Miss Lyon and Miss Whitman in the department of chemistry. Her zeal in the laboratory has been exceeded only by her enthusiasm in botany. A graduate of 1851, she has spent most of her life in the service of the seminary. To her efforts largely it owes its botanical collection, its botanical garden, and a conservatory that is only a promise of the one she has in her eye. With the exception of Mrs. Mary A. (Hurd) Foster, one

THE ORNITHOLOGICAL ALCOVE.
Lyman Williston Hall.

of the domestic superintendents, she is the only member of the seminary that ever saw Miss Lyon. Although crowned with the almond blossom she is deterred from her favorite pursuits by neither distance nor difficulty. But after a year's absence for the purpose of studying the flora of the Hawaiian Islands she returns to the seminary just too late to furnish the reader with a report of her acquisitions there.

ALUMNÆ ASSOCIATIONS.

Mention has been made of Miss Lyon's calling together the Holyoke alumnæ present at the meeting of the American Board in Norwich, in 1842. There have always been informal gatherings at the seminary anniversary. In 1853, fifteen out of sixteen classes, and in 1862 all the twenty-five classes were thus represented. In 1871 seventy graduates from twenty-four classes discussed methods of rendering aid to their Alma Mater. The result was the formation of a "National Association of Holyoke Alumnæ," whose "object is to promote the prosperity of the seminary." It was formed in New Haven, Connecticut, by the alumnæ present at the meeting of the American Board in 1872. Holyoke gatherings continue in connection with the annual meetings of that body, but since its first year the Association has held its annual meetings at the seminary, anniversary week.

The New Haven Association was formed a few months before the National, to which it then became auxiliary. Branches have since been formed in the following places, in the order named: Springfield, Chicago, Worcester, Danvers, Hawaiian Islands, Boston, Philadelphia, New York, Hartford, and Greenfield. Any pupil though not a graduate may join; membership in a branch constitutes membership in the National Association. One dollar is the annual fee.

Gifts from alumnæ, amounting in value to several thousand dollars, not only aided in building Williston Hall, but have done much toward filling its cabinets and furnishing each department it represents, besides contributing in many ways to the comfort and enjoyment of the household. A grateful record is kept of every gift and donor.

Thanks are due to Miss Sarah A. Clarke, of the class of '79, for collecting and arranging, in addition to the facts here given, full statistics respecting the membership and gifts of the several associations. Her report is a paper of great value for future use.

At its annual meeting in 1885, the Association took the first steps toward obtaining endowment funds. It is hoped that the Mary Lyon Fund, for endowing the principal's chair, will be completed in time to be a jubilee gift to Alma Mater, and will speedily be followed by funds for other departments.

Mrs. Moses Smith, of Detroit, Michigan, is president of the National Association; Miss Louise F. Cowles, recording secretary; and Miss Sarah H. Melvin, treasurer. By the constitution, the two officers last named must be filled by "members of the faculty resident at the seminary." The list of vice-presidents includes, with others, the presidents of the several branches.

The membership of the two organizations being practically the same, it has been proposed to unite the Memorandum Society and the Alumnæ Association. But since the work of the one is that of the mother following with interest her absent daughters, and of the other that of the daughters seeking to aid the mother, it has been thought hitherto that the two objects could be accomplished better by co-operation than by union.

CHAPTER XXI.

INSTITUTIONS MODELED AFTER MOUNT HOLYOKE SEMINARY.

MOUNT Holyoke Seminary is itself the embodiment of certain important ideas. The results of its work are seen not only in the impress it stamps upon its students, and in its own recognized position, but also in the movement that has originated so many other institutions for the higher education of women and which is one of the most important developments in the life of our country and in the history of our race. Miss Lyon's prayers and plans embraced the world and were for all time. As the years roll by, her prayers are being answered and her faith justified in the outgrowths of her work. The system she developed is not only adapted to secure the highest order of intellect for the service of Christ in our own land, but has proved to be no less useful in training mind and forming character in other lands. In a greater or less degree its peculiar features characterize many institutions at home and abroad. Their history would fill volumes, but space can be taken for only the following sketches prepared mainly from material furnished for the purpose.

To the frequent question, what are the essential features of Mount Holyoke Seminary, it is answered: That it may realize its ideal—the greatest usefulness—it must combine thorough mental training with careful religious culture; and put this education within reach of the class most likely to be benefited by it and to use it for the good of the world. In order to do this it must be devoted to the service of Christ; "furnished with every advantage which the state of education will allow;" and able to put its charges low.

The first call for another Holyoke school came from Persia in 1843, through Rev. Dr. Perkins, who visited the seminary and asked for a teacher to establish a similar institution for the Nestorians. Out of many candidates, Miss Fidelia Fiske was selected to go.

When mission work began in Oroomiah only one among the thousands of Nestorian women could read. The men opposed the education of women. Through the efforts chiefly of Mrs. Grant, a few girls had been gathered into a day school. Little, however, could be done for them till they were separated from the degradation of their surroundings. In order to bring them under the influence of a Christian home the mission provided accommodation and support for six boarding pupils. Although a boys' school had been easily gathered, there was much doubt whether parents would allow their daughters to enter a boarding school. But Miss Fiske in faith prepared her house, and on the day appointed, Mar Yohanan brought two little girls, saying: "They are your daughters; no man shall take them out of your hand. Now, you begin Mount Holy Oke in Persia." In a few months the six boarders were secured and the work of the seminary began. Twenty-five boarders gladdened the heart of Miss Fiske the second year, and although the buildings were enlarged, there were more applicants the third year than they could contain. In training these untutored girls, Miss Fiske's great abilities were taxed to their utmost. She was mother, housekeeper, and teacher, bearing cheerfully a load of care that would have crushed a less loving heart.

In the early days of the school the Bible was the only text-book and the teaching was oral; but Miss Fiske was a teacher of rare power, and the effect of the constant presentation of the Bible as the rule of life was soon apparent. Meanwhile a change was taking place in the community. The visits of the missionaries were welcomed. The preaching of the Gospel found eager listeners among the women as well as the men. The winter of 1846 was made memorable by the first of those remarkable revivals with which that mission has been blessed. The work began on the first Monday of the year. The pupils of both schools were bowed down under a deep sense of sin. Importunate prayer was heard on every side, some even making places in the woodshed for private devotion. In the two schools, fifty were soon rejoicing in Christ. Then the parents and friends came in to learn what this "awakening" meant, and to be brought themselves under the power of the Spirit. Miss Fiske writes: "I had often ten or fifteen women to spend the night with us. The young converts were full of zeal in leading them to Christ. The voice of prayer was often heard all through the night." When the pupils separated for vacation the divine leaven was carried into many families and in spite of persecution the work advanced.

As Miss Fiske's cares increased she turned to her Alma Mater for help, and one like-minded, Miss Mary Susan Rice, gladly responded, reaching Oroomiah in November, 1847. Even then she found the school a "miniature Holyoke." For more than twenty years she labored in the school with untiring zeal, spending her vacations, as did Miss Fiske, in seeking out and instructing the women of remote villages and of the mountains. The discomforts of touring were alleviated here and there by finding Christian homes presided over by pupils who welcomed their teachers with holy joy. Years later a missionary says of these homes, "They are lighthouses in the great, dark sea of iniquity which covers the mountains."

The history of the school is one of steady growth, despite many hindrances. The course of study was gradually extended, the Bible always retaining the chief place. The girls were trained to do a work as teachers, wives, and mothers, which should sanctify their nation. February came to be designated as the month of blessing, and the last weeks of the winter term prepared the girls to be very useful in their village homes. The last evening of one term a pupil stopped at the teacher's desk as she passed out from family worship, to make a special request. "If you please, will you allow our class to pray together just as long as we wish to-night? We are going home to-morrow; we can sleep then." Permission granted, the class separated only with the dawn.

When, after nearly sixteen years of such sowing and reaping, Miss Fiske was forced by failing health to return to America, ninety-three women sat down with her at the table of our Lord. And these were only a part of those whom God had blessed through her labors.

Miss Aura J. Beach, of the class of '59, went to assist Miss Rice after Miss Fiske left; but though fitted in every way save that of health, she was able to remain only two years in the field.

Miss Lucy M. Wright, now Mrs. Mitchell, gave a year to the service; and Mrs. Sarah J. (Foster) Rhea, after the death of her lamented husband, continued his work for Persia by entering with her wonted enthusiasm into Miss Rice's work in the seminary. She thus describes a revival occurring at that time: "Miss Rice was a true daughter of Mary Lyon; the Lord was honored and he honored the school by his special presence. After devotions one morning, the seven seniors, all Christians, were seated for the recitation in theology. The lesson that day was about Christ, the sinless Saviour, dying for the world. First rose Miriam. She announced her subject, gave the divisions, and dwelt on the perfections of this wonderful Redeemer. All at once her voice trembled, broke beyond control, and she sat down in tears. The next rose and told what Christ was in heaven, and what he left to die for his enemies; suddenly she felt the personal application, and left her theme unfinished. As she dropped into her seat, another followed, developing a kindred thought, and broke down in the same way. The next rose from deep study of Isaiah's vision which told how He should die, brought as a lamb to the slaughter, wounded for our transgressions, bruised for our iniquities. Then, conscious of her own portion in the story she was telling, and unable to proceed, she covered her face and sat down. Next to her sat Yasmin, who, after long resistance of the Spirit, had been brought into the kingdom two years before. She had lived eighteen years in sin, but had been led to Christ by these words which she found one day: 'Though your sins be as scarlet, they shall be as white as snow; though they be red like crimson, they shall be as wool.' She recited next, telling how God loved the world by giving his only begotten Son, and she said solemnly, 'If the dying love of the Son of God fails to melt a human heart, there is no power in the universe to move it!' and her own heart was so moved, she, too, was unable to say more. We fell on our knees to worship, and the recitation was turned into prayer."

Later in the same year, Mrs. Rhea and Miss Rice returned to America, the latter exhausted by the labors of many years. The seminary was left in charge of Miss Jennie Dean, of Michigan, who still continues at her post.

The preparatory and seminary courses now occupy six years, and a year of post-graduate study is offered in addition.

The widespread influence of the school was strikingly manifested at the recent jubilee of the Nestorian Mission, by the presence of nearly eight hundred intelligent Christian women. Cheering reports were brought of pupils of the seminary, who in the darkness and isolation of the mountain villages still retain their love for the Bible, studying it constantly and teaching it to those about them.

CHEROKEE SEMINARY.—In 1851, John Ross, a chief of the Cherokee nation, after visiting many schools, fixed upon the Holyoke system as best adapted to develop character and advance Christian education among his people. Brick buildings were erected in Tahlequah [now in Indian Territory], and a three years' course of study arranged, embracing the higher English branches, Latin, and vocal music. In its early years the school was under the care of Mount Holyoke graduates. Its students are filling important and responsible places, being a power for good among their people.

THE WESTERN SEMINARY, in Oxford, Ohio, was first suggested by a few earnest Christians living in Oxford, led by Mr. and Mrs. Daniel Tenney. A beautiful site of thirty acres, now increased to sixty-five, adjoining the town, was given by James Fisher. Gabriel Tichenor and family, of Walnut Hills, gave the first five thousand dollars; other generous gifts followed. Trustees were appointed in July, 1853, and the building began. The enterprise was laid before the principal and teachers of Mount Holyoke Seminary, with the request that they would foster this western daughter, assist in making her a counterpart of the mother institution, and select the first corps of teachers from their own ranks. Of these, Miss Helen Peabody, who had been associated with Mary Lyon as pupil and teacher, was elected principal. The seminary was dedicated September 20, 1855.

The history of its trials is wonderful; more wonderful is the way God has brought it through them all, victorious through that which has made her daughters invincible,—God within. Three thousand pupils have received instruction, and four hundred have graduated. Fifty are foreign missionaries, while others are doing service no less valuable in the home field. A large number are doing good work as teachers, and five have entered the medical profession. God has yearly visited this school with his saving grace; hundreds have been converted, and many others strengthened in the Christian life.

The school year of 1859-60 was brought to a sudden close by the fire of January 14th, originating in a defective flue. There was no excitement, no confusion, though one of the pupils had a narrow escape. The doors of the Oxford Female College were hospitably thrown open to receive the homeless family. The senior class completed their studies in the house of Mr. James Fisher, near the seminary grounds, under the instruction of Miss Peabody and Miss McCabe, the graduating exercises occurring in May. An interesting feature of this occasion was the fitting up of a box of documents to be placed in the corner-stone of the second building. These words, written by members of the graduating class, were included, and are significant as history: "The Western Female Seminary; Christ himself the chief corner-stone. It has nothing to fear but that it may not know its duty or may fail to do it. May the pillar of cloud which guarded the door of the tabernacle rest ever at the going in of our beloved seminary." The new building was dedicated in May, 1862. In the preceding year the seminary had received a permanent fund of twenty thousand dollars, the bequest of one of its earliest friends, Gabriel Tichenor, the income to be applied to the salaries of its instructors. In the list of teachers for that year we find a new name, Miss Emily Jessup. Affectionate allusions are made to her invalid condition as well as to the great value of her instruction in class room and in religious meetings.

Only a few hours before the building was consumed by fire the second time, April 7, 1871, earnest prayer had been offered that when the proposed stone porch should be erected, the completed building might be more entirely consecrated as an abiding place for the King of kings. The next morning, a pupil told Miss Peabody that her books were lying in front of the ruins, and on the top "Müller's Life of Trust." "I do believe," she added, "it means God is going to give us back our seminary, just as he gave Müller the orphan houses."

The Lord did, indeed, prompt many hearts to send their offerings unsolicited. The trustees were not long in deciding to rebuild. By the middle of October they had provided an attractive and commodious home, and on the day before Thanksgiving, trustees, alumnæ, and friends gathered for the dedication service.

In alluding to the almost miraculous strength given the young ladies who rescued Miss Jessup from the flames on that fearful night, it was remarked: "I believe it was God who did it, while he permits those girls to have the pleasure of thinking it was their work. So," said the speaker, "I believe it was God who wrought all these marvelous things we give thanks for to-day, but he allows these friends to think they are doing the work in his name."

"It has been our greatest desire," says Miss Peabody, "that our school should be Christ's; that each pupil should be one of the King's daughters; that every room should be an abiding place for the Holy Spirit." We cannot recount the many ways in which these prayers have been answered. In the winter of 1871, the religious interest began early in the term and deepened during the week of prayer; but the pentecostal day was not till later in the year, when nearly the whole family confessed Christ. Of several years following, equally interesting records might be made. Temporal blessings were not denied in connection with the spiritual. Improvements have been made in the buildings and furnishings, as well as in the course of study. New apparatus has been obtained, and large additions have been made to the library, which now contains over four thousand well chosen volumes.

The approximate value of seminary property is one hundred and twenty thousand dollars, and the invested funds amount to thirty-four thousand dollars, having been largely increased within the last ten years by the munificent gifts of the late honored treasurer of the board of trustees, Mr. Preserved Smith, of Dayton, Ohio. A students' fund is available to those looking forward to Christian work. One feature of this seminary—a fundamental one in its establishment—is the

aid rendered the daughters of home missionaries. The
original charge for board and tuition was only sixty
dollars a year. In 1887, it is one hundred and seventy,
which covers gas-light and heating by steam. The
course of study, embracing four years, is somewhat less
extended than that of Mount Holyoke, thus admitting
younger pupils. Lectures by professors from different
institutions supplement instruction in the class room.
The department of art is well furnished with models of
a standard character, and special attention is paid to
music. A systematic course of Bible study extends
through the four years. The principles of education
inculcated by Mary Lyon are faithfully reproduced, and
the spirit with which she infused the Holyoke system is
felt by every Oxford pupil.

LAKE ERIE SEMINARY, in Painesville, Ohio, was the
successor of Willoughby Seminary, founded in 1847,
and discontinued by the burning of the building in
1856. That seminary was noted for thorough scholarship and decided religious influence. The number of
pupils averaged two hundred in the later years, with a
graduating class of fourteen. A preparatory department was a necessity, but the regular course of study
was nearly the same as at Mount Holyoke Seminary,
where most of its teachers had been educated, among
them two nieces of Miss Lyon. Miss Roxena B. Tenney
was principal till 1854, and was succeeded by Miss
Marilla Houghton and Miss Julia M. Tolman. After
the building was burned, citizens of Painesville made
liberal offers, and the trustees voted to locate the seminary there, and upon a somewhat different plan. A
new board of trustees was formed; Rev. Roswell Hawks,
who had been invited to Ohio to advise in the matter,
was appointed to solicit subscriptions. The corner-
stone of the building was laid July 4, 1857. A Holyoke feature which had not been possible at Willoughby, was the family life, all the students being
gathered under one roof and sharing in the domestic

duties. The grounds of the seminary comprised fourteen acres, including a grove of oaks and chestnuts, under whose shade the anniversary exercises have been held through successive years. The building, four stories high and one hundred and eighty feet long, accommodates one hundred and fifty students. The school was opened in September, 1859, by three teachers and four other graduates from Mount Holyoke Seminary, Miss Lydia A. Sessions being principal. There were one hundred and twenty-seven pupils and two graduates at the close of the first year. Notwithstanding the civil war, the seminary prospered during the administration of Miss Sessions, which continued till 1866, when Miss Anna C. Edwards was appointed principal. Upon her return to Mount Holyoke Seminary in 1868, she was succeeded by Miss Mary Evans,—like her predecessors, a highly esteemed Holyoke teacher. In 1878, Miss Luette P. Bentley, one of its own graduates, was appointed associate principal.

Since its beginning, the seminary has had more than three thousand students. Two hundred and fifty-one have graduated, one hundred and sixty of whom have engaged in teaching. Forty have occupied important positions in high schools and seminaries. Five have studied medicine and one has graduated from a law school. Twelve graduates and former students have engaged in foreign mission work, and seven are missionary teachers in the South and in Utah. These numbers would be increased if statistics could be obtained in regard to members of the school, not graduates.

The standard of admission to the regular course of study has been gradually raised, and the course, occupying four years, now includes the same amount of Latin and mathematics as at Mount Holyoke, and a liberal course in history, literature, philosophy, and the natural sciences,—laboratory work being required in chemistry, physics, and botany. The domestic department prospers. Labor-saving appliances have been introduced, courses of lectures in domestic economy have been delivered, and great value is placed on this important part of a symmetrical education.

As to its financial history, a heavy debt was incurred at the outset, which did not diminish till, chiefly by a gift of ten thousand dollars from Hon. Reuben Hitchcock, president of the board of trustees, it was wholly removed in 1871. His benevolent work was seconded by his colleagues in the board, according to their ability, some of them taking charge of seminary affairs so thoroughly that the services of a steward have been dispensed with for eighteen years. The largest improvement in the building was the erection, in 1876, of a wing, seventy by forty feet and three stories high. The approximate value of seminary property including grounds, buildings, and apparatus is one hundred and fifty thousand dollars.

The invested funds of the seminary amount to thirty-two thousand dollars. The rate of board and tuition, at first ninety dollars exclusive of fuel and lights, and at present two hundred dollars, includes all advantages except private lessons in music and painting. The number of pupils in 1887 is one hundred and thirty.

Its religious history bears witness to the faithfulness of a covenant-keeping God. The means of grace, so richly blessed in the development of Christian character in the mother school, have availed here: the morning half hour in the chapel, for praise and prayer and exposition of the Word; the silent half hours when each student is alone in her quiet room; the prayer meetings; the systematic course of Bible study through the four years; and the days set apart for prayer, especially the day of prayer for colleges. Not a school year has passed without special religious interest, and, although it is not easy to sum up results, it is safe to say that at the end of each year the number not professedly Christian has averaged less than one-tenth of the whole. Of the two hundred and fifty graduates only five are known to have been without a Christian hope.

The seminary has been undenominational. Each denomination has its own missionary society, meeting monthly for the study of its special field at home and abroad, and contributing to its own missionary board. On the first Sunday of the month these societies unite in a "monthly concert," where reports and papers of general interest are presented.

In MARSOVAN, TURKEY, Miss Ann Eliza Fritcher, of the class of '57 and a Holyoke teacher, has been for more than twenty years in charge of a seminary on the Holyoke principles which is a center of blessing in educating Christian women.

MILLS SEMINARY AND COLLEGE in California was founded by Rev. Dr. and Mrs. C. T. Mills, "to do for the far West what Mount Holyoke Seminary does for the East." Dr. Mills had been president of Batticotta Seminary, Ceylon, and of Oahu College, Hawaiian Islands. Mrs. Susan (Tolman) Mills graduated in 1845, taught in the seminary until 1848, and after her marriage was closely associated with her husband in his labors. They began their work in California at Benicia, in 1865. Prosperous from the beginning, the school was removed in 1871 to Brooklyn, Alameda County, where a better location for a permanent institution was secured, commodious buildings were erected in a lovely park of eighty-five acres, and excellent facilities provided. For several years more the school was carried on as a private enterprise, and gained the confidence of the Pacific coast by the cultured graduates who went out from it year by year. In 1877 the founders committed the seminary so largely the fruit of their own prayers and labors, to the hands of trustees, though they still remained at the head of it until the death of Dr. Mills in 1884, when the trustees placed it in charge of Mrs. Mills. In 1885 the institution was re-incorporated and a college curriculum was added. It was the aim of the founders " to establish a Christian school on a permanent basis, that like Mount Holyoke it should depend on no individual life, but should become a wellspring of blessing to California and the world." This hope has already been realized in large measure and will, we doubt not, be more fully realized in the future history of the college. Including the class of 1886 more than sixteen hundred young ladies have been enrolled as pupils. Three hundred and twelve, representing different states and territories as well as the Sandwich Islands, Alaska, British Columbia, and Mexico, have graduated.

The charter of the MICHIGAN SEMINARY at Kalamazoo specifies that it shall be "essentially modeled after the Mount Holyoke Seminary, founded by Mary Lyon." The buildings stand in a beautiful oak grove overlooking the city, the river, and adjacent country,—the whole forming a wide and beautiful landscape. The school was opened in January, 1867. Miss Jeanette Fisher, a Holyoke graduate of 1859, was appointed principal, and held that position for twelve years. More than a score of the teachers have been Mount Holyoke graduates. It has been repeatedly blessed with revivals, and a large proportion of its more than one hundred graduates have become active Christian workers. They are found in all parts of our country and some are foreign missionaries. The school has labored at times under financial difficulties, but it has done and is doing excellent work. It is now free from debt. Its facilities for instruction are much increased and its friends anticipate for it a prosperous future.

THE MOUNT HOLYOKE SEMINARY at Bitlis, Turkey, is in one of the "darkest, loneliest corners of the world, hidden behind the mountains of Kurdistan." It is under the care of Misses Charlotte E. and Mary A. C. Ely, of the class of '61. They began their work in 1868, after Mrs. Knapp had taught a few pupils and prepared the way. Few parents cared to have their daughters instructed, but the success of the mission soon increased the number. A good building was erected, additions have been made to it, and in these attractive rooms two hundred pupils have received a Christian education.

In the summer, tents are pitched on a mountain about three miles from the city. In this healthful retreat the teachers pursue their work during the hot season. Sixteen pupils have finished the course of study, receiving a diploma from "The Mount Holyoke Seminary, Bitlis." Like that of Alma Mater it bears the inscription, "That our daughters may be as corner-stones, polished after the similitude of a palace."

The study of the Bible, the prayer meetings, and missionary meetings are all powerful influences in the education of these Koordish girls. Two precious revivals have been enjoyed, and the Misses Ely have the "unspeakable delight of seeing many of their pupils consecrated to the Lord, and carrying his gospel to darker regions beyond."

THE HUGUENOT SEMINARY at Wellington, Cape Colony, the oldest Holyoke offshoot in South Africa, was founded for European colonists, and opened January 19, 1874.

The Rev. Andrew Murray, pastor of the Dutch Reformed church there, on reading the "Life of Mary Lyon," said to Mrs. Murray, "Such a school is just what we need for our own daughters and for the daughters of our people." He interested the churches of the Colony in the subject and asked their prayers for God's blessing upon the undertaking. With a liberal hand they supplied the funds required. Letters were sent to Mount Holyoke Seminary, asking for a graduate to take charge of the school. Their faith that the request would be granted appeared in their sending money for her passage before receiving a reply to their appeal.

Miss Abbie P. Ferguson, of the class of '56, and Miss Anna E. Bliss, of '62, responded to the call. They went out scarcely knowing whither they went, but believing that they were led by the hand of God. When the news reached Africa that two were coming, a little company gathered about the open letters and gave thanks to Him who had given double what was asked. On their arrival they received a hearty welcome from the clergymen of the Dutch Reformed church, then assembled in synod, and from the people of Wellington.

The seminary received the name of Huguenot, in honor of the French refugees, who came to Cape Colony from Holland, whither they had fled from persecution after the revocation of the "Edict of Nantes." Many of their descendants are still living in the Colony, exerting a great power in every good cause.

A large building, surrounded by pleasant grounds, was purchased. Its location reminded the new-comers of their Alma Mater, the ground sloping down to another "Stony Brook," with "Prospect Hill" rising beyond. The two institutions are similar in the arrangements of the home and the plan of the school. A course of study was adopted much the same as that of Mount Holyoke in 1837. Above all there was an earnest desire that the school should be eminently Christian. The "Life of Mary Lyon," translated into Dutch, had been read with much interest, and forty pupils assembled for the opening besides day scholars, who formed a separate school in the village. There was a spirit of hearty co-operation among the students from the first. Before the end of the term every pupil expressed the hope that she had accepted Christ as her Saviour.

The demand for greater advantages has been met from year to year. Teachers have gone from the United States and from different parts of Europe, and some of the graduates have remained to teach. A normal department has been added, supplying the need for trained teachers, and improving the village schools. Two new buildings have been erected, one on either side of the first, so that more than one hundred boarders can be accommodated, while one hundred and fifty children are connected with the normal department.

Through the liberality of A. Lyman Williston, Esq., of Northampton, Massachusetts, an observatory built upon the grounds contains the telescope used at Mount Holyoke Seminary until the erection of its present observatory. A hall, with rooms on the first floor for art and science, the gift of Hon. E. A. Goodnow, of Worcester, Massachusetts, promises to double the usefulness of the institution.

In December, 1878, the first class graduated. From that time the seminary has been on a sure foundation, and though the standard of attainment has been raised a class has graduated each year. The students find many opportunities for practical Christian work among the colored population, in gathering the children into Sabbath-school, and holding prayer meetings in the cottages. Many go forth from the seminary to be teachers; others occupy positions of responsibility and are found in cultivated homes in all parts of the Colony. During the first twelve years, ten became missionaries.

The labors of Rev. Andrew Murray have been blessed to both pupils and teachers. Each year there have been special seasons of revival.

The influence of Mount Holyoke Seminary in South Africa is not limited to the Huguenot Seminary; nor would the history of the latter be complete without an allusion to the progress of Christian education through kindred institutions in other parts of the land. It was no sooner established than there was a demand for others of like stamp. In November, 1874, Miss Juliette Gilson, of the class of '68, arrived in Stellenbosch, Cape Colony, in response to an application from Rev. J. Neethling of that place, for a Holyoke graduate. She took charge at once of a work of great promise; and opened the Bloemhof Seminary in 1875. The plan of instruction, which at first was necessarily limited to elementary subjects, now embraces English and Dutch literature, German, French, Latin, geometry, evidences of Christianity, and natural theology. The first class graduated in 1881. Many besides graduates have gone forth to be blessings in the land. Like other Holyoke seminaries the Bloemhof is not local in its interests. Out of one hundred and eighty-four pupils in its first five years, twenty-four were from Stellenbosch, forty-four from Cape Town and vicinity, while the remainder came from more than thirty different places in the Colony and the regions beyond, including Diamond Fields, Orange Free State, Basuto Land, Natal, and the Transvaal.

Rev. William Murray, of Worcester, was the third pastor to send for teachers. As his brother had done before, he forwarded passage money, and began preparation in confident expectation of success; nor was he disappointed. Miss Ellen Smith, of the class of '73, and her sister, Miss Annie Smith, reached the Cape late in 1875. When the seminary building in Worcester was completed—April, 1876—all the American teachers then in the Colony assembled for its dedication, and to confer regarding the interests of Christian education in South Africa. It was decided to ask at once for six more teachers from America.

In the same year, Miss Helen Murray, who had been two years at the Huguenot Seminary, began an important work at Graaf Reinet, taking charge of the Midland Seminary, till the arrival of Miss Thayer, of the class of '60. On Miss Thayer's return to America in 1880, Miss Murray again became principal. During the first term of the school there was a revival almost as remarkable as that of the first term at the Huguenot Seminary.

In 1877, Rev Andrew Murray, and his brother, Rev. Charles Murray, returning from America, were accompanied by ten teachers. One of the ten—Miss Martha Newton, of the class of '75—opened a prosperous school at Swellendam.

At Pretoria, in the Transvaal, a school was opened in 1877 by Miss Susan M. Clary, who had been fourteen years a teacher at Mount Holyoke Seminary. The journey from Cape Town, one thousand miles by sea and five hundred overland, was difficult and trying. A large number of pupils gathered around Miss Clary, catching her enthusiasm. A severe attack of pneumonia resulted in consumption, and in less than a year she entered into rest, rejoicing that it had been given her to do something for Africa. Of the forty or more teachers who have gone from the United States to these schools more than twenty were Holyoke pupils. Applications have generally come through Rev. Andrew Murray, of Wellington, and the selection has been made under the direction of the principal of Mount Holyoke Seminary and Mrs. H. B. Allen, of Meriden, Connecticut.

The reading of the memoirs of Miss Lyon and Miss Fiske has been followed by efforts to establish schools on the Holyoke plan in England and in France.

A SPANISH HOLYOKE SEMINARY is under the care of Mrs. Alice (Gordon) Gulick, of the class of '67. From the translation of a letter from Sr. Dn. Cipriano Tornos, a Madrid pastor, we make the following extracts:—

"Passing through San Sebastian we improved with pleasure the happy opportunity of being present at the examinations then being held. They were the following: 'Reading, Writing, Arithmetic, Grammar, History of Spain, Geography, Universal History, Reading Music at Sight, Singing, Exercises upon the Piano and Organ, Spanish Literature, French-English Grammar, English Literature, Bible History, Book-keeping, Theory of Teaching, Gymnastic Exercises, Drawing, and Embroidery, as well as Plain Sewing.' Perhaps reading this list one would fear to see accomplished here the Spanish adage, ' He who tries to do much accomplishes little.' No! In no wise! The young girl who at last obtains her diploma attesting that she has finished the studies here taught is able not only to talk about them, but can dedicate herself to the work of teaching them. We have seen here proved what is the current opinion in Spain, that in the Evangelical schools there is more and better teaching than in others.

"The number of pupils in the different departments during the year is eighty-two. Of these three finished the course of study and we had the pleasure and privilege of presenting them their diplomas. Two of them expect to teach. All this we have seen ourselves, no one has told it to us. But we have seen more, which has surprised us beyond measure. At the same time that these pupils prepared for their examinations, they found time to prepare for a brilliant musical soirée, which took place the following day at night."

The writer enumerates the "classical pieces played and sung," closing with "the magnificent Hallelujah Chorus by Farmer," and congratulates the director of the school, teachers, and scholars, "and lastly the American Board, which with such generosity sustains this school in San Sebastian," and continues:—

"It now remains for us to give some general information about the aforesaid school. It was founded in Santander in the year 1876, and was transferred to San Sebastian in 1881. The object is not only to prepare teachers for usefulness but to give an ample and solid education to all those young girls who are able to attend.

"Especial care is given to educate the scholars in the life of a well organized home. They are taught to do for themselves to-day, what to-morrow they will have to do in their own homes. That is to say, they are taught to be good housekeepers, not mere señoritas of the drawing room.

"They pay according to their ability. Scholars of all ages are admitted, for as the whole house of five stories is given up to the school, the scholars are cared for according to their age, forming but one family.

"In regard to religious instruction it is understood to be essentially and eminently Biblical. Every morning before breakfast there is family worship in the chapel; sessions of study, meals, etc., are preceded by prayer.

"Every evening the children have a prayer meeting before retiring, and on Sunday the older girls have a meeting for mutual edification and prayer. Thursday evening of each week there is public worship in the chapel. On Sunday there are two meetings with sermon at 11 A. M. and 8 P. M., besides the Sabbath-school at three o'clock. Besides all this there are daily Bible classes for the whole school in sections, according to age.

"This is what we have truly seen and what we have learned regarding this school."

In Japan a translation of the "Life of Mary Lyon" has been much sought and widely read by men as well as women. Not to mention others, a school on the Holyoke plan was opened in 1885 at Kanazawa by a graduate of the seminary at Oxford, Ohio. Beginning with thirty-two pupils it has more applicants than can be received. A letter dated December, 1886, speaking of the desire expressed by officials of a neighboring city for a similar school, illustrates the favor which the higher education for women is gaining in Japan.

Other Holyoke offshoots in mission stations and elsewhere are doing similar work and sending forth pupils to multiply good influences. To the work of Mary Lyon, Mr. D. L. Moody traces the establishment of his seminaries at Northfield and Mount Hermon.

WELLESLEY COLLEGE, opened in 1875, has a close connection with Mount Holyoke Seminary through its founder, Henry F. Durant, Esq., a trustee of both institutions, who diligently studied the Holyoke system that he might embody its essential features in his ideal of a Christian college for women. The character and history of this prosperous and truly Christian college, its wide influence, and the great and good work it has already accomplished in the twelve years of its existence are too well known to require description here.

ALBERT LEA COLLEGE at Albert Lea, Minnesota, belongs to the Presbyterian churches of that state. It was opened in September, 1885, and at the dedication of the building was emphatically declared a Holyoke school, whose object should be the training of young women for Christian work, especially to an interest in missions.

The domestic system of Mount Holyoke and many of its school and family regulations have been adopted. Its principal is Miss Laura C. Watson, of the class of '71. who writes: " Rich spiritual blessings have already been granted, and though only two years old. it has three scholarships for those who have declared their intention of devoting their lives to missionary work. Its literary curriculum is as high as that of the best colleges for women in New England. The number of students in 1887 is sixty."

As another has said: "The seminary like a banyan tree spreads abroad its branches and takes root in many a foreign soil, while the mother trunk grows only the more stately and strong beside the same 'river of water' where it was so wisely planted at first."

CHAPTER XXII.

CATALOGUE OF OFFICERS OF MOUNT HOLYOKE SEMINARY, OF ANNIVERSARY SPEAKERS, AND OF PASTORS.

TRUSTEES OF MOUNT HOLYOKE SEMINARY.

NOTE.—Of those still in office the present residence is given; of others, that at the time of election.

	Elected.	Died or Resigned.
Hon. William Bowdoin, South Hadley Falls,	1836	res. 1856
Rev. John Todd, D. D., Northampton,	1836	res. 1836
Rev. Joseph D. Condit, South Hadley,	1836	d. 1847
Hon. David Choate, Essex,	1836	res. 1843
Hon. Samuel Williston, Easthampton,	1836	res. 1836
Rev. William Tyler, South Hadley Falls,	1836	res. 1856
Rev. Roswell Hawks, Cummington,	1836	d. 1870
Hon. Joseph Avery, Conway,	1836	d. 1855
Andrew W. Porter, Esq., Monson,	1836	d. 1877
Rev. Heman Humphrey, D. D., Amherst,	1836	res. 1846
Rev. Edward Hitchcock, D. D., LL. D., Amherst,	1836	d. 1864
Hon. Daniel Safford, Boston,	1837	d. 1856
Hon. Samuel Williston, Easthampton,	1839	res. 1862
Rev. E. Y. Swift, Northampton,	1847	res. 1874
Rev. Samuel Harris, D. D., LL. D., Conway,	1848	res. 1856
Rev. Edward N. Kirk, D. D., Boston,	1856	d. 1874
Hon. Edward Southworth, West Springfield,	1856	d. 1869
Abner Kingman, Esq., Boston,	1856	d. 1880
Austin Rice, Esq., Conway,	1858	d. 1880
Rev. Theron H. Hawks, D. D., Springfield,	1858	res. 1861
Rev. Hiram Mead, D. D., South Hadley,	1859	res. 1873
Rev. William S. Tyler, D. D., LL. D., Amherst,	1862	
Ariel Parish, M. A., Springfield,	1864	res. 1865
Sidney E. Bridgman, Esq., Northampton,	1865	
Rev. John M. Greene, D. D., Lowell,	1866	res. 1875
Henry F. Durant, Esq., Boston,	1867	res. 1879
A. Lyman Williston, M. A., Northampton,	1867	
Rev. Nathaniel G. Clark, D. D., LL. D., Boston,	1868	
Hon. William Claflin, LL. D., Boston,	1869	
Edward Hitchcock, M. A., M. D., Amherst,	1869	
Rev. Julius H. Seelye, D. D., LL. D., Amherst,	1872	
Hon. Edmund H. Sawyer, Easthampton,	1873	d. 1879

OFFICERS.

	Elected.	Died or Resigned.
Rev. John R. Herrick, D. D., South Hadley,	1875	res. 1879
Francis A. Walker, LL. D., New Haven, Conn.,	1876	res. 1886
Rev. John L. R. Trask, Lawrence,	1879	
Henry D. Hyde, Esq., Boston,	1880	
Charles A. Young, Ph. D., LL. D., Princeton, N. J.,	1880	
Rev. William De Loss Love, D. D., South Hadley,	1881	
Rev. William M. Taylor, D. D., LL. D., New York,	1881	
John H. Southworth, Esq., Springfield,	1881	
G. Henry Whitcomb, M. A., Worcester,	1881	
Miss Elizabeth Blanchard, *ex officio*,	1884	
Mrs. A. Lyman Williston, Northampton,	1884	
Mrs. Helen M. (French) Gulliver, Somerville,	1886	

PRESIDENTS OF THE BOARD OF TRUSTEES.

Rev. John Todd, D. D.,	1836	res. 1836
Rev. William Tyler,	1837	res. 1838
Rev. Roswell Hawks,	1838	res. 1858
Rev. Edward N. Kirk, D. D.,	1858	d. 1874
Rev. William S. Tyler, D. D., LL. D.,	1874	

SECRETARIES.

Rev. Joseph D. Condit,	1836	d. 1847
Rev. E. Y. Swift,	1848	res. 1859
Rev. Hiram Mead, D. D.,	1859	res. 1869
Rev. John M. Greene, D. D.,	1869	res. 1874
Edward Hitchcock, M. A., M. D.,	1874	res. 1880
Rev. John L. R. Trask,	1880	res. 1882
Rev. William De Loss Love, D. D.,	1882	

TREASURERS.

Samuel Williston,	1836	res. 1836
Hon. William Bowdoin,	1836	res. 1856
Hon. Samuel Williston,	1856	res. 1862
Andrew W. Porter, Esq.,	1862	res. 1873
A. L. Williston, M. A.,	1873	

PRINCIPALS.

Abbreviations in this and following lists: m, **married**; d, **died**.

	Class.	Term.
Mary Lyon,	—	1837–49
d. 1849.		
Mary C. Whitman,	'39	1849–50
m. Morton Eddy, 1851; d. 1875.		
Mary W. Chapin, acting principal,	'43	1850–52
Mary W. Chapin,		1852–65
m. Claudius B. Pease, 1865.		

350   *MOUNT HOLYOKE SEMINARY.*

	Class.	Term.
Sophia D. (Hazen) Stoddard, acting principal,	'41	1865–67
m. William H. Stoddard, 1867.		
Helen M. French,	'57	1867–72
m. Lemuel Gulliver, 1872.		
Julia E. Ward,	'57	1872–83
Elizabeth Blanchard,	'58	1883–

ASSOCIATE PRINCIPALS.

Eunice Caldwell,	—	1837–38
m. Rev. John P. Cowles, 1838.		
Abigail Moore,	'38	1842–46
m. Rev. Ebenezer Burgess, 1846; d. 1853.		
Mary C. Whitman,	'39	1842–49
Sophia D. Hazen,	'41	1849–50
m. Rev. David T. Stoddard, 1851.		
Sophia Spofford,	'46	1852–55
Emily Jessup,	'47	1855–62
Julia M. Tolman,	'48	1858–60
m. Lucius A. Tolman, 1862, d. 1871.		
Catharine Hopkins,	'54	1860–65
d. 1865.		
Mary Ellis,	'55	1867–72
Julia E. Ward,	'57	1867–72
Elizabeth Blanchard,	'58	1872–83
Anna C. Edwards,	'59	1872–

TEACHERS.

Lucy M. Ainsworth,	'49	1849–51
m. T. D. Strong, M. D., 1852.		
Paulina Avery,	'50	1852–53
m. Rev. O. L. Woodford, 1856; d. 1858.		
Laura W. Ayer,	'54	1855–56
d. 1860.		
Hannah O. Bailey,	'39	1844–45
m. Rev. H. O. Howland, 1845.		
Elizabeth D. Ballantine,	'57	1860–69
m. Rev. Charles Harding, 1869.		
Elizabeth M. Bardwell,	'66	1866–
Mary E. Barker,	'46	1846–46
d. 1846.		
H. Augusta Belcher,	'60	1860–61
m. S. J. S. Rogers, M. D., 1865; d. 1877.		
Mary J. Belcher,	—	1855–58
Mary A. Berry,	'85	1885–
Elizabeth Blanchard,	'58	1858–
Mary E. Blodgett,	'70	1875–76

TEACHERS.

	Class.	Term.
Emily W. S. Bowdoin,	'49	1849-50
m. James Armour, 1855.		
Sarah Bowen,	'64	1870-75
Susan Bowen,	'64	1868-75
m. David S. Jordan, 1875; d. 1885.		
Ellen P. Bowers (absent six years),	'58	1861-
Martha E. E. Bradford,	'69	1871-76
m. Oscar Gilchrist, M. D., 1876.		
Mary C. Bradford,	'71	1880-
Ella T. Bray,	'83	1883-85
m. Harris Graham, M. D., 1885.		
Mary A. Brigham,	'48	1855-58
Sarah Brigham,	'38	1838-39
m. Rev. C. B. Kittredge, 1840; d. 1871.		
Mary P. Bronson,	'58	1858-59
m. T. S. Bridgman, 1860; d. 1865.		
Marie F. Browne,	'44	1844-45
d. 1847.		
Mary Q. Brown,	'49	1849-52
Susan N. Brown,	'51	1851-52
Laura A. Buckingham,	'73	1873-75
Elizabeth Burt,	'51	1855-57
Martha R. Chapin,	'42	1842-46
m. Rev. Allen Hazen, D. D., 1846; d. 1884.		
Mary W. Chapin,	'43	1843-65
m. Claudius B. Pease, 1865.		
S. Elizabeth Chapin,	'54	1854-55
M. Elizabeth Childs,	'56	1859-70
m. H. Danforth Perry, 1871.		
Cornelia M. Clapp,	'71	1872-
Caroline W. Clark,	'59	1861-66
m. Harding Woods, 1867.		
Martha J. Clark,	'86	1886-
Susan M. Clary,	'63	1863-77
d. 1878.		
Harriet M. Cooley,	'53	1853-56
m. Rev. Amos H. Coolidge, 1856.		
Lydia G. (Bailey) (Rogers) Cordley,	'45	1881-82
Louise F. Cowles,	'66	1867-
Lucy M. Curtis,	'44	1844-49
d. 1849.		
Mary S. Cutler,	'75	1876-78
Annie Dearborn,	'65	1866-71
m. Rev. Cyrus Richardson, 1871.		
Martha C. Dole,	'41	1843-44
Susan S. Driver,	'67	1867-68
Elizabeth Earle,	'60	1860-67
m. Rev. George F. Magoun, D. D., 1870.		

	Class.	Term.
Anna C. Edwards (absent five years),	'59	1859–
Lucy J. Ellis,	'62	1875–76
Mary Ellis,	'55	1855–72
Sarah A. Emmons,	'53	1855–56
m. Hon. Thomas Spooner, 1856.		
Mary A. Evans,	'60	1860–68
Fidelia Fiske,	'42	1842–43
Also (part of the time),	"	1859–64
d. 1864.		
Rebecca W. Fiske,	'46	1846–49
m. Rev. Burdett Hart, 1849.		
Helen C. Flint,	'80	1881–82
Mary M. Foote,	'47	1858–59
Nancy A. Foote,	'44	1844–45
m. Rev. Edward Webb, 1845.		
Mary J. Forbes,	'65	1865–66
m. Rev. Daniel C. Green, 1869.		
Helen M. French,	'57	1857–72
m. Lemuel Gulliver, 1872.		
Ann Eliza Fritcher,	'57	1859–63
Sarah A. Gilbert,	'49	1851–53
m. Henry Anderson, 1853; d. 1856.		
Helena F. Giles,	'71	1876–77
Anna C. Gilman,	'49	1849–50
m. Rev. Charles D. Lothrop, 1854; d. 1864.		
Julia A. Goodhue,	'64	1866–68
m. Prof. Hoyt Trowbridge, 1871.		
Alice W. Gordon,	'67	1868–70
m. Alvah B. Kittredge, 1870.		
m. Rev. William Gulick, 1871.		
Mary E. Graves,	'44	1845–46
m. Sylvanus Miller, 1850.		
Adaline E. Green,	'67	1871–
Susan F. Hawks,	'42	1842–43
m. Rev. Charles A. Williams, 1855.		
Eliza C. Haskell,	'56	1859–64
m. Rev. Edward S. Frisbie, D. D., 1864.		
Mary Haynes,	'64	1864–68
m. Rev. John W. Lane, 1868.		
Frances M. Hazen,	'63	1865–
Sophia D. Hazen,	'41	1844–50
m. Rev. David T. Stoddard, 1851.		
Persis D. Hewitt,	'76	1878–79
Harriet A. Hinsdale,	'44	1855–57
m. Rev. Henry L. Hubbell, 1863.		
Elizabeth P. Hodgdon,	'69	1870–75
m. Rev. Lester H. Elliot, 1875.		

TEACHERS.

	Class.	Term.
Amanda A. Hodgman,	—	1837–39
m. —— Nourse, 1852; d. 1854.		
Lucy J. Holmes (absent five years),	'58	1858–81
Also,	"	1886–
Ann M. Hollister,	'45	1845–47
m. Rev. A. B. Campbell, 1851.		
Anna M. Hood,	'69	1869–72
m. Lucius E. Hall, 1874.		
Henrietta E. Hooker,	'73	1873–
Catharine Hopkins,	'54	1854–65
d. 1865.		
Ada L. Howard,	'58	1858–61
Mary H. Humphrey,	'43	1844–46
m. Silas Ames, M. D., 1856; d. 1859.		
Helen Humphrey,	'39	1839–41
m. Albert A. Palmer, Esq., 1845.		
m. William H. Stoddard, 1852; d. 1860.		
Ellen Hunt,	'54	1854–55
m. Jonathan Bacon, 1858.		
Myra M. Jenkins,	'64	1864–66
m. Martin L. Mead, M. D., 1867.		
Sabrina Jennings,	'43	1847–48
Emily Jessup,	'47	1847–60
Harriet Johnson,	'46	1848–52
m. Rev. R. M. Loughridge, 1853.		
Mary L. Judd,	'76	1886–
Marcia A. Keith,	'83	1885–
Marietta Kies (absent three years),	'81	1881–
Mary A. Kimball,	'53	1855–56
m. Jacob P. Palmer, 1859; d. 1882.		
Catharine E. Lee,	'54	1854–56
d. 1874.		
Jane E. Lemassena,	'57	1857–58
Sarah D. Locke,	'59	1859–68
m. Rev. John M. Stow, 1868.		
Louisa A. Long,	'53	1853–54
m. J. P. Woodbury, 1857.		
m. Wm. H. Woodbury, 1862.		
Eliza A. Lyon,	'56	1856–58
Lucy T. Lyon,	'40	1841–46
m. Rev. Edward C. Lord, 1846; d. 1853.		
Isabella G. Mack,	'75	1875–86
Margaret Mann,	'42	1844–48
m. Rev. Thomas O. Rice, 1850.		
Maria E. Mason,	'49	1849–50
m. Rev. M. K. Cross, 1852; d. 1855.		
Catharine McKeen,	—	1852–56
d. 1858.		

354   *MOUNT HOLYOKE SEMINARY.*

	Class.	Term.
Phebe F. McKeen, . . .	—	1853–56
d. 1880.		
Harriette A. Melvin, ...	'56	1856–58
Sarah H. Melvin, ...	'62	1870–
Caroline H. Merrick, ...	'49	1849–50
Mary B. Metcalf, ...	'46	1847–48
m. Rev. Edward Chester, 1854.		
Abigail Moore, ...	'38	1838–46
m. Rev. Ebenezer Burgess, 1846; d. 1853.		
Ann R. Mowry, ...	'42	1842–43
m. Rev. Jeremy W. Tuck, 1845.		
Mary A. Munson (absent five years), . .	'48	1848–57
m. S. S. Burton, Esq., 1857; d. 1881.		
Mary J. Murdock, ...	'50	1850–51
m. George R. Gold, 1857.		
Martha L. Newcomb, ...	'48	1850–51
Lura E. Newhall, ...	'64	1864–66
m. Jay Phetteplace, 1869.		
m. George F. Greer, 1879.		
m. T. R. Levitt, 1886.		
Hannah Noble, ...	'58	1861–
Mary O. Nutting, ...	'52	1870–
Olive L. Parmelee, ...	'61	1862–68
m. Rev. Alpheus N. Andrus, 1875.		
Anna A. Parsons, ...	'70	1876–79
Ellen C. Parsons, ...	'63	1883–85
Roxana R. Parsons, ...	'41	1841–45
m. Caleb Green, M. D., 1845; d. 1885.		
Sarah P. Parsons, ...	'66	1869–71
Elizabeth K. Peabody, ...	—	1866–67
Helen Peabody, ...	'48	1848–53
Mary F. Phinney, ...	'52	1852–55
m. Henry K. Whiton, Esq., 1855; d. 1865.		
Clara C. Pond, ...	'57	1858–61
m. Rev. W. F. Williams, 1866.		
Catharine A. Porter, ...	'44	1844–45
m. Rev. F. H. Pitkin, 1845.		
m. Rev. Addison Lyman, 1847.		
Elizabeth B. Prentiss, ...	'62	1866–
Lucinda T. Prescott, ...	'53	1853–56
Harriet E. Reed, ...	'64	1865–67
m. Austin E. Messenger, 1876.		
Susan Reed, ...	'39	1839–44
m. Rev. W. Howland, 1845.		
Lois W. Rice, ...	'45	1853–54
m. Thomas E. Hale, 1854; d. 1873.		
Lydia G. (Bailey) Rogers, ...	'45	1850–52
m. Rev. C. M. Cordley, 1852.		

TEACHERS.

	Class.	Term.
Laura A. Rose,	—	1883–
Elizabeth I. Samuel,	'80	1880–84
Helen M. Savage,	'68	1868–70
m. Rev. Albert Ball, 1870.		
Emily A. Scott,	'52	1852–53
m. Charles W. Cleveland, 1859 ; d. 1860.		
Hannah C. Scott,	'43	1848–49
m. Francis E. Clarke, 1858.		
Martha C. Scott,	'45	1845–55
m. D. O. Dickinson, 1856.		
Harriet E. Sessions (absent three years),	'56	1857–
Lydia A. Sessions,	'56	1856–59
m. Rev. W. W. Woodworth, 1866.		
Lydia W. Shattuck,	'51	1851–
Lillie L. Sherman,	'80	1880–84
S. Effie Smith,	'86	1886–
Eliza Smith,	'51	1851–52
m. Rev. Jesse L. Howell, 1853 ; d. 1871.		
Mary W. Smith,	—	1837–38
m. R. A. Severance, M. D., 1838 ; d. 1844.		
Matilda W. Smith,	'58	1859–60
m. Wm. A. Magill, 1860.		
Minerva Smith,	'54	1855–58
m. E. Hazen, 1858 ; d. 1864.		
Sophie A. Smith,	—	1884–86
m. Rev. Arthur W. Burt, 1887.		
Sophia Spofford,	'46	1851–55
Also,	"	1871–72
M. Ella Spooner,	'72	1872–84
Sarah A. Start (absent two years),	'52	1852–59
d. 1872.		
Mary F. Stearns,	'53	1857–59
m. Rev. Augustine Root, 1860 ; d. 1877.		
Clara F. Stevens,	'81	1881–
Louise P. Stevens,	'59	1859–61
Mary M. Stevens,	'42	1842–48
Sophia D. (Hazen) Stoddard,	'41	1864–67
m. William H. Stoddard, 1867.		
Sarah D. (Locke) Stow (part of the time),	'59	1877–
Calista A. Streeter,	'57	1857–59
m. Orlando Mason, 1859.		
Abbie L. Sweetser,	'74	1875–
Gertrude Sykes,	'53	1853–58
m. Rev. Quincy Blakely, 1858.		
Esther E. (Munsell) Thompson,	'59	1869–70
Persis G. Thurston,	'45	1845–47
m. Rev. T. E. Taylor, 1847.		

	Class.	Term.
Elizabeth Titcomb,	'50	1850-53
m. Benjamin V. Abbott, Esq., 1853.		
Mary Titcomb,	'50	1850-56
Jane C. Tolman,	'51	1858-64
Julia M. Tolman (absent two years),	'48	1851-60
m. Lucius A. Tolman, 1862; d. 1871.		
Susan L. Tolman,	'45	1845-48
m. Rev. Cyrus T. Mills, D. D., 1848.		
Sarah H. Torrey,	'39	1839-43
m. Rev. Henry Eddy, 1843; d. 1885.		
Mary C. Townsend,	'62	1867-80
Frances V. Turner,	'58	1858-59
d. 1862.		
Jessie Usher,	'57	1860-61
Susan M. Waite,	'55	1856-58
m. Rev. Edward P. Thwing, 1859.		
Adelia C. Walker,	'51	1852-53
Julia E. Ward,	'57	1857-83
Delia H. Warner,	'74	1875-78
d. 1879.		
Frances E. Washburn,	'69	1870-72
Ann R. Webster,	'42	1843-45
m. Rev. Horace Eaton, D. D., 1845.		
Aurilla P. Wellman,	—	1848-51
m. A. F. Hitchcock, M. D., 1851; d. 1862.		
Annie M. Wells,	'67	1870-74
Caroline Wentworth,	'53	1853-55
m. Edward M. Morse, 1860.		
Mary C. Whitman,	'39	1839-50
m. Morton Eddy, 1851; d. 1875.		
Caroline L. White,	'71	1872-74
Adeline H. Willcox,	'52	1853-54
m. Chandler Richards, Esq., 1859.		
Emily S. Wilson,	'61	1861-63
Clara W. Wood,	'73	1873-
Persis C. Woods,	'38	1838-39
m. Rev. George C. Curtis, D. D., 1848.		
Amelia C. Woodward,	'58	1859-60
m. Orran P. Truesdell, 1864.		
Sarah A. Worden,	—	1883-
Catharine A. Wright,	'42	1842-45
m. John M. Brewster, M. D., 1846; d. 1851.		
Mary E. Yale,	'48	1849-50
d. 1852.		
Caroline R. Yates,	'51	1851-52
m. Rev. Smith B. Goodenow, 1853.		

TEACHERS.

PHYSICIANS AND TEACHERS OF PHYSIOLOGY.

	Class.	Term.
Mary A. B. Homer, M. D.,	—	1860–64
m. Samuel D. Arnold.		
Emily N. Belden, M. D.,	—	1864–68
m. Albert N. Taylor.		
m. John McCabe; d. 1886.		
Lucy M. Southmayd, M. D.,	—	1868–70
m. Lucius Garvin, M. D.		
Emma H. Callender, M. D.,	—	1869–73
d. 1878.		
Charlotte W. Ford, M. D.,	'62	1873–74
Olive J. Emerson, M. D.,	'65	1874–76
m. Rev. Horatio Morrow, 1876.		
Adaline D. H. Kelsey, M. D.,	'68	1876–78
Adelaide A. Richardson, M. D.,	—	1878–82
d. 1885.		
Fanny G. Heron, M. D.,	—	1882–85
Juliet E. Marchant, M. D.,	—	1885–86
Elizabeth M. Peck, M. D.,	'76	1886–

VOCAL MUSIC.

Frances M. Atwood,	'39	1838–39
m. E. M. Cowles, 1841; d. 1852.		
Deborah E. N. Bates,	'40	1839–40
Amelia F. Dickinson,	'44	1841–44
m. Edward D. Bangs, 1844.		
Harriet Hawes,	'48	1844–48
Sarah F. Woodhull,	—	1848–49
Emily W. S. Bowdoin,	'49	1849–50
m. James Armour, 1855.		
Lucy C. Mills,	'52	1850–52
m. S. T. Brooks, M. D., 1856.		
Martha A. Bailey,	'53	1852–53
Catharine E. Lee,	'54	1853–54
d. 1874.		
Elizabeth W. Shepard,	'57	1854–55
m. G. W. Neill, 1860.		
Charlotte Morgan,	'56	1855–56
m. Julian Pomeroy.		
Lucinda D. Hodge,	'60	1857–60
d. 1861.		
Fanny M. Hidden,	'62	1860–62
m. Rev. B. G. Page, 1865; d. 1870.		
Eliza Wilder,	—	1862–66
m. Rev. Henry M. Holmes, 1867.		
Almeda N. Tirrell,	'66	1866–69
m. J. Gilman McAllister, M. D., 1869.		

358   *MOUNT HOLYOKE SEMINARY.*

	Class.	Term.
Mary P. Burgess,	'69	1869–71
Abby A. Wilder,	—	1871–72
m. Daniel Davis, 1872.		
Emma A. Ide,	—	1872–73
m. Rev. John A. Cruzan.		
Annie S. Wilson,	—	1873–75
d. 1881.		

VOCAL AND INSTRUMENTAL MUSIC.

	Class.	Term.
Charlotte M. Steele,	—	1875–86
Ada J. MacVicar,	—	1878–80
m. —— Carmen.		
Eva F. Pike,	—	1881–85
Florence E. Balch,	—	1885–
Florence Grinnell,	—	1886–

FRENCH.

	Class.	Term.
Abigail Moore,	'38	1837–46
J. A. Lucie Robinson,	'48	1846–48
m. Rev. Edwin R. Beach, 1861.		
Mary A. Munson,	'48	1848–50
m. S. S. Burton, Esq., 1857; d. 1881.		
Eleanor Kevney,	—	1850–51
Gertrude de Bruyn Kops,	—	1851–52
d. 1852.		
Sarah J. Gillette,	—	1853–54
Mary E. Peabody,	'61	1859–61
d. 1870.		
Lydia Richards,	'66	1863–66

FRENCH AND GERMAN.

	Class.	Term.
Caroline de Maupassant,	—	1872–73
Valerie Dietz,	—	1873–75
Margarethe Vitzhum von Eckstadt,	—	1875–85
Marie Gylam,	—	1885–86
Anna E. Engelhardt,	—	1886–

ASSISTANT PUPILS.

	Term.
Abigail Moore,	1837–38
Persis C. Woods,	1837–38
m. Rev. George C. Curtis, D. D., 1848.	
Susan Reed,	1837–39
m. Rev. Wm. W. Howland, 1845.	
Martha A. Leach,	1838–39
m. Rev. Wm. S. Curtis, D. D., 1845.	
Lucy T. Lyon,	1839–40
m. Rev. Edward C. Lord, 1846; d. 1853.	

OFFICERS.

	Term.
Rachel Blanchard,	1839–40
m. Joseph Pool, 1840.	
m. Abner P. Nash, 1854; d. 1859.	
Maria K. Whitney,	1839–40
Sarah M. Paine,	1840–41
Mary M. Stevens,	1840–42
Julia Hyde,	1840–42
m. Rev. Edward Clark, 1844.	
Lucy M. Curtis,	1842–43
d. 1849.	
Caroline Avery,	1844–45
m. Norton A. Halbert, Esq., 1850; d. 1880.	
Caroline H. Merrick,	1848–49
Mary A. Burt,	1855–56
d. 1877.	
Lillie L. Sherman,	1878–80
Adelaide S. Phillips,	1880–81
m. Warren Steele, 1882.	
Ella T. Bray,	1883–84
m. Harris Graham, M. D., 1885.	

SUPERINTENDENTS OF DOMESTIC DEPARTMENT.

Miss Emily Bridge,	1841–45
m. John Smith, Esq., 1847.	
Mrs. Mary K. Carroll,	1859–65
Mrs. Mary A. Foster,	1864–
Mrs. Mary K. Carroll,	1868–73
d. 1887.	
Miss Sarah A. Thayer,	1873–74
Mrs. Harriet G. Dutton,	1874–82
Mrs. R. L. Wright,	1882–

STEWARDS.

Ira Hyde,	1838–44
Rev. Roswell Hawks,	1844–55
John H. P. Chapin,	1855–66
Ithiel Lawrence,	1866–85
David E. Phillips,	1885–86
Lewis H. Porter,	1886–

ANNIVERSARY SPEAKERS AND SUBJECTS.

1838. Rev. Joel Hawes, D. D.
1839. Rev. Rufus Anderson, D. D., LL. D.; *Education of the Female Sex.*
1840. Rev. Mark Hopkins, D. D., LL. D.; *Female Education.*
1841. Rev. Bela B. Edwards, D. D.; *Female Education.*
1842. Rev. Edward Hitchcock, D. D., LL. D.; *Waste of Mind.*

1843. Rev. Lyman Beecher, D. D.; *Education.*
1844. Rev. Edward N. Kirk, D. D.; *The Greatness of the Human Soul.*
1845. Rev. Joel Hawes, D. D.; *The Virtuous Woman.*
1846. Rev. J. B. Condit, D. D.; *Christian Love the Essential Element of Moral Power.*
1847. Rev. Heman Humphrey, D. D.; *The Systematic Cultivation of the Mind.*
1848. Rev. Edward Beecher, D. D.; *Christian Experience Indispensable to the Normal Development and Culture of all the Powers, and to a Complete Education.*
1849. Rev. Edward Hitchcock, D. D., LL. D.; *A Chapter in the Book of Providence.*
1850. Rev. William C. Fowler, LL. D.; *Taste.*
1851. Rev. A. L. Stone, D. D.; *The Mission of Woman.*
1852. Rev. Barnas Sears, D. D., LL. D.
1853. Rev. Emerson Davis, D. D.; *Female Education.*
1854. Rev. Edwards A. Park, D. D., LL. D.; *Conscience.*
1855. Rev. Theodore D. Woolsey, D. D., LL. D.; *Woman under the Feudal System.*
1856. Rev. Samuel W. Fisher, D. D.; *John Calvin and John Wesley.*
1857. Rev. George Shepard, D. D.; *Reading.*
1858. Rev. William A. Stearns, D. D., LL. D.; *Female Education.*
1859. Rev. Austin Phelps, D. D.; *The Oneness of God in Nature and Revelation.*
1860. Rev. Roswell D. Hitchcock, D. D., LL. D.; *The Laws of Civilization.*
1861. Rev. Asa D. Smith, D. D., LL. D.; *Self-possession.*
1862. Rev. R. S. Storrs, D. D., LL. D.; *The Country and the Times.*
1863. Rev. John L. Hart, LL. D.; *Attention.*
1864. Rev. Leonard Swain, D. D.; *Puritan Education.*
1865. Hon. B. G. Northrop, LL. D.; *Religious Education.*
1866. Rev. N. G. Clark, D. D., LL. D.; *The Moral Element, the Vital Force in Literature.*
1867. Rev. George N. Boardman, D. D.; *Importance of Large Institutions of Learning.*
1868. Rev. V. D. Reed, D. D.; *Importance of the Christian Scriptures in Education.*
1869. Col. C. G. Baylor; *Woman's Work.*
1870. Julius H. Seelye, D. D., LL. D.; *The Power of Ideas.*
1871. Rev. Jacob M. Manning, D. D.; *Goethe's Writings and Character.*
1872. Rev. H. D. Kitchell, D. D.; *The Virtuous Woman.*
1873. Rev. William S. Tyler, D. D., LL. D.; *The Higher Education of Woman.*
1874. Rev. Samuel Harris, D. D., LL. D.; *Religion Essential to Culture.*
1875. Rev. Julius H. Seelye, D. D., LL. D.; *The True Groundwork of Education.*
1876. Hon. Alexander H. Bullock, LL. D.; *The Centennial Situation of Woman.*

1877. Rev. M. B. Riddle, D. D.; *Bible Study in our Higher Institutions of Learning.*
1878. Rev. Thomas P. Field, D. D.; *Taste: Its Culture and Applications.*
1879. Rev. Theodore L. Cuyler, D. D.; *The Pathways of Life.*
1880. Rev. S. E. Herrick, D. D.; *The Librations of Thought.*
1881. Hon. Albion W. Tourgee, LL. D.; *Excessive Activity of Americans.*
1882. Rev. William M. Taylor, D. D., LL. D.; *Reading.*
1883. Rev. Joseph T. Duryea, D. D.; *Development of Mind.*
1884. Rev. John H. Vincent, D. D., LL. D.; *Woman's Sphere.*
1885. Rev. F. D. Huntington, D. D.; *The Law of Social Life and Woman as Its Executor.*
1886. Hon. Henry L. Dawes; *Now and Then. Opportunities of Modern Life.*
1887. Rev. Wm. S. Tyler, D. D., LL. D.; *The Semi-Centennial Address.*

PASTORS.

In the public worship of the Sabbath the seminary unites with the First Congregational Church in South Hadley, the only church in the village, and pays from its treasury a certain sum toward parish expenses. The pastor occasionally conducts religious services in the seminary, is usually one of the board of trustees, and always takes a warm interest in this large part—about one-third—of his congregation.

July 8, 1835, six months from the day that South Hadley was fixed upon for the location of the seminary, Rev. Joseph D. Condit was installed pastor of the church. He died September 19, 1847. His successors have been:—

Rev. Thomas Laurie, D. D., installed June 7, 1848; dismissed Feb. 17, 1851.
Rev. Eliphalet Y. Swift, installed Nov. 3, 1852; dismissed Feb. 9, 1858.
Rev. Hiram Mead, D. D., ordained Sept. 29, 1858; dismissed Nov. 19, 1867.
Rev. John M. Greene, D. D., installed Feb. 26, 1868; dismissed May 31, 1870.
Rev. J. Henry Bliss, installed Jan. 11, 1871; dismissed May 1, 1873.
Rev. John R. Herrick, D. D., installed April 16, 1874; dismissed April 16, 1878.
Rev. William De Loss Love, D. D., commenced as supply about Sept. 1, 1878; installed May 7, 1879.

INDEX.

Academies, origin of, 6; incorporation of Dummer, Abbot, Bradford, Ipswich, Leicester, and Westford, 6, 7; Adams, 7, 9, 29, 30, 31; Sanderson and Amherst, Mary Lyon at, 16, 18, 34.
"Accounts balanced," 298.
Adams, Jacob, bequest of, 29.
Afternoon exercises, 106, 125, 131, 135.
Agassiz, Louis, 238.
Age of students, 56, 72, 94, 103, 139, 281.
Aid from the state refused, 224; granted, 225.
Allen, Mrs. H. B., 344.
Allen, Nathan, M. D., 229.
Allen, Peter, 84; Levi, 89.
Alumnæ, use of the word, 318; at work, 159, 189, 190, 278, 318-326; teaching, 111, 158, 159, 190, 318; missionaries, 24, 159, 173, 189, 204, 205, 220, 278, 318, 319; see *Teachers at Mount Holyoke*; for what not fitted, 289; gifts from, 241, 268, 269, 270, 274, 323, 324, 325; associations, 325; testimony of, 117-137, 288-317; acknowledgment to, vi.; biographical record of, vi.; mortality of, 229; daughters of, 312.
American Board, 23, 24, 162, 176, 204, 208, 254, 262, 310, 325, 345.
Anderson school of natural history, 238.
Anderson, Rufus, D. D., LL. D., 68, 200, 205, 215; quoted, 103, 138, 140, 177, 204, 350.
Andrew, Hon. John A., LL. D., 195.
Anniversary, the first, 100-102; addresses quoted from, 139, 140; speakers and subjects, list of, 359.
Applicants, many, 89, 94, 138, 186, 192.
Arnold, Thomas, D. D., and Mary Lyon compared, 26.
Art, study of, 244, 255.
Artesian well, 257.
Association, Franklin County, action of, 30, 32.
Association, Mass. General, action of, 42.
Associations, Alumnæ, 325.
Avery, Hon. Joseph, 57, 61, 63, 64, 348.

Bailey, Prof., 228.
Baldwin, Rev. Theron, letter to, 94.
Ballantine, Elizabeth D., Mrs. Harding, 209, 210, 231, 319, 350.
Bancroft, Mrs. Ruth (Washburn), gift of, 268.
Banister, Mrs. Wm. B., see *Grant, Z. P.*, 166, 204, 243; letters to, 166, 175.
Bardwell, Elizabeth, 260, 273, 350.
Barker, Mary E., death of, 176; 350.
Beach, Aura J., 205, 319, 330.
Beecher, Catharine E., Hartford Seminary, 8; objections of to Miss Lyon's plans, 43; answered, 44-46.
Beecher, Lyman, D. D., 47.
Bemis, Edward W., Ph. D., 271.
Benevolence, Christian, of donors, 40, 62, 66, 91; of teachers, 91; of pupils, 24, 69, 91, 174; training in, 113, 161, 230.
Bentley, Luette P., 336.
Bible, the, and girls, in earlier times, 3, 15; in course of study, 27, 30, 105, 146, 282; made paramount, 105-7, 149, 161, 201, 231, 301; Bible and science, 22, 147, 300, 301; Miss Lyon's, 132, 133.
Bierstadt, gift of, 251.
Billings, Hammett, 227.
Blanchard, Elizabeth, associate principal, 235, 350; 236, 255, 268; principal, 269, 277; trustee, 278, 349; administration of, 268-283.
Blanpied, Prof., 248.
Bliss, Anna E., 323, 340.
Bliss, Rev. Daniel, 211.
Bloemhof Seminary, sketch of, 342.
Boat-house, 247-272.
Boston Journal, quoted, 85.
Boston Recorder, 41.
Boswell, Charles, fund for library, 280.
Botanical garden, 272, 273, 324.
Bowdoin, Hon. William, 39, 61, 64, 348.
Bowen, Susan, Mrs. Jordan, 238, 351.
Bowers, Ellen P., 271, 351.
Bowker, Mrs. Albert, 23.
Bray, Ella T., Mrs. Graham, 278, 319, 351.
Bread-making, 93.

Bronson, Mary P., Mrs. Bridgman, 302, 303, 351.
Brooks, S. D., M. D., 121, 180.
Browne, Sarah K., 90.
Bruce, Rev. Henry J., gifts of, 324.
Buckland, birthplace of Mary Lyon, 3; missionary society of, 23; school at, 22, 30, 34, 41.
Buildings; erection of first, 62-77; site, 40, 64, 65; corner-stone, 65; discouragements, 64, 65, 66, 74, 76; described, 84, 87; why not larger, 65; furnished, 66, 74, 75, 77, 88; dedicated, 90; first addition, 142-3, 153; opposed, 142; north wing, 191, 192; gymnasium and south wing, 195, 196, 222, 224; library, 193, 226, 227, 270; improvements in, 178, 191, 247; fire escape, 248; elevator, 257, 258; cost of buildings, 65, 144, 227, 251, 260; see *Williston Hall*; *Observatory*.
Bullock, Hon. A. H., LL. D., 225, 360.
Burgess, Mrs. Mary (Grant), 33.
Burgess, see *Moore, Abigail*; 175, 177, 178, 321; letters to, 172, 174, 176.
Burnham, Samuel, 200.
Burritt, Elijah H., 17.
Burt, Rev. Arthur W., 278.
Burt, Mrs. Sophie A. Smith, 278, 355.
Bushnell, Rev. Albert, 207.

Caldwell, Eunice, 67, 85, principal Wheaton Seminary, 70; gift of, 85; associate principal at Mount Holyoke, 69, 72; 350; quoted, 101; see *Cowles*.
Calisthenics, practice in, 104, 146, 192.
Catalogue of trustees of Mount Holyoke Seminary, 348; teachers, 349-359; superintendents of domestic department, 359; stewards, 359; anniversary speakers, 359-361; pastors, 361.
Catalogues, of early schools, 16; of Mt. Holyoke, statistics from annual, 284-287; from memorandum, 318.
Catechism, Shorter, in early schools, 3.
Catholic schools for girls, 10, 11, 72.
Chadbourne, Paul A., D. D., LL. D., 197.
Chamberlain, George, 268.
Chapin, John H. P., 191, 198, 211, 359.
Chapin, Martha R., Mrs. Hazen, 174, 320, 351.
Chapin, Mary W., 186; acting principal, 187, 349; principal, 188, 349; 190, 192, 203, 204, 209-213, 215, 217-219, 315, 351; administration of, 187-219; tribute of trustees, 187, 218; marriage of, 220; father of, 187; see *Pease*.
Character of first pupils, 69, 88, 94, 99, 143.

Charter of Mt. Holyoke Seminary, 11; copy of, 61.
Choate, Hon. David, 39, 61, 63, 348.
Churchill, Rev. J. Wesley, 228.
Circulars quoted: of general committee, 54; to candidates, 55; for furnishings, 66; responses to, 74; "General View," 71; prospectus, 72; in 1840, 142.
Clapp, Cornelia M., 250, 274, 351.
Clark, Alvan, 259.
Clark, N. G., D.D., LL.D., 220, 348, 360.
Clark, Wm. S., Ph. D., LL.D., 248.
Clarke, Sarah A., 326.
Clary, Susan M., 265, 323, 343, 351.
Clinton, Gov. of New York, 8.
Colleges, why founded, 2; why not for girls, 2; number in 1802, 2; in 1836, 11: Amherst, how founded, 34; history of, quoted, 80: Georgia, 256: Harvard, founded, 1, 11; life at in 1830, 85; curriculum, 86; not an "experiment," 141; growth, 149: Mills, 338; Oxford, O., 333: Smith, 239: Vassar, 256: Wellesley, origin of, 239, 239; 256, 263, 346: Yale, 149.
Condit, Rev. Joseph D., aid of, 39, 61, 75, 175; addresses, 90, 161, 165; trustee, 348; pastor, 361; death, 176; Mrs. Condit, 161.
Connecticut valley, historic, 79; geology of, 79, 83.
Conn. Valley Botanical Society, 247.
Conscience trained, 111, 112, 289, 291, 296.
Conversions, 22, 106, 150, 152, 153, 156-158, 169, 170, 175, 178, 185, 202, 206, 207, 212, 214, 218, 230, 263, 264, 265, 276, 277; labor for, see *Revivals*.
Converts trained, 107, 130, 171, 305.
Cook, Rev. Joseph, 249.
"Cornelius" Rutherford, 275, 304.
Corner-stone, laying of, 65, 246.
Cost of buildings, see *Buildings*.
Course of study, at Byfield, 27; Buckland and Derry, 30; Ipswich, 30, 56; Harvard, 86; Mt. Holyoke, 56, 104, 105, 140; compared with Ipswich, 145; outline of in 1837, 145; in 1887, 281; progressive, 56, 105, 145, 148, 196, 197, 228, 236, 244, 267, 270; additions, 147, 148, 244, 255; see *Bible*; *Languages*; *Electives*; *Lecture courses*.
Cowles, Mrs. Henry, 9.
Cowles, Mrs. John P., see *Caldwell*; 158, 204; quoted, 68, 71.
Cowles, Louise F., 274, 326, 351.
Criticism, notes of, 124.
Cupola, 144.
Curtis, Lucy M., 185, 351.

INDEX. 365

Dana, Daniel, D. D., 39, 63.
Dascomb, Mrs., 9.
Day of prayer, 108, 214; described, 151, 231, 264; see *Fast-days*.
Dean, Jennie, 331.
Death of a student, 129.
Debt of the seminary removed, 222, 225.
Derry, N. H.; see *Academies, Adams*.
Dickinson, Emma F., gift of, 268.
Dickinson, Mr., gift of telescope, 193.
Diploma, first given to women, 30; of Mt. Holyoke Seminary, 101; of seminary at Bitlis, 340.
Domestic department; suggested, 35; why adopted, 36, 91, 97; design of, 72, 98; misunderstood, 97, 98; an experiment, 73, 96; superintendent hard to find, 92, 97; system complicated, 92, 96, 135; described, 95; modified, 99, 275; **failure** predicted, 91, 98, 99; result of experiment, 85, 95; advantages of, 91, 96, 97, 115, 230, 295, 304; testimony of alumnæ, 290, 292-295, 304; theory tested, 91-99.
Donors, see *Avery; Bancroft; Bierstadt; Boswell; Bruce; Caldwell; Dickinson, E. E.; Dickinson, Mr.; Durant; Durant, Mrs.; Eldridge; Fairbank; Goodnow; Gridley; Hazeltine; Kendall; Kingman; Maynard; Merriam; Porter; Porter, Mrs.; Safford; Sawyer, Mrs.; Wheaton Seminary; Williston, A. L.; Williston, Mrs. A. L.; Williston, Samuel*; see also *Alumnæ; Finances; Gifts*.
Drawing, taught, 148, 268, 274, 283.
Dress, attention to, 118, 124, 126, 127.
Dummer, Wm., 6; see *Academies*.
Durant, H. F., labors of, 219, 230; benefactions, 223, 226, 228; trustee, 348; plans for Wellesley, 230, 239; death, 263.
Durant, Mrs., gift of, 194, 226, 228, 263.
Dwight, Elihu, 88, and John, 89.
Dwight, Mrs. Mary (Billings), 33.
Dwight place, the, 268, 269, 274.

Eaton, Prof. A., 17.
Eddy, Mrs. Morton, see *Whitman*; 243, 244, 272.
Eddy, Zachary, D. D., 208.
Education, decline in, 3; effect of Revolution, 4; for girls in 18th century, 2-7; first law for improvement of, 8; first legislative aid, 8; superficial, 10; prejudice against, 7, 8, 10, 139; higher for women, an experiment, 139, 140, 141, 204; change in public opinion, 236, 239; see *Schools; Theory*.

"Education of American girls," quoted, 229.
Education Society, first for women, 43.
Edwards, Anna C., associate principal, 236, 250, 252, 268, 269, 272, 336, 350, 352.
Edwards, Bela B., D. D., 49, 138, 140, 359.
Edwards, Jonathan, 14, 150.
Eldridge, John B., bequest of, 280.
Elective courses, 282.
Elective studies, 271, 274, 284; in second and third years, 282.
Elevator, 144, 257, 258, 272.
Ellis, Mary, associate principal, 221, 227, 234, 235, 350, 352.
Ely, Charlotte E. and Mary A. C., 320, 339.
Ely Volume, The, on Missions and Science, 323.
Emerson, Olive J., M. D., Mrs. Morrow, 265, 320, 357.
Emerson, Rev. Joseph, views of, 27-29; work of, 9, 17, 21, 27, 28; influence of, 18, 28, 29.
Emerson, Prof. Ralph, 43.
Endowment funds needed, 281, 326.
Evans, Mary, 225, 336, 352.
Expenses at girls' schools, 9, 10, 33, 34; at Mt. Holyoke first year, 72, 73; next sixteen years, 100; increased, 191, 226; table, 284; see *Finances; Tuition*.
Experiment, the seminary an, 139, 140, 204.
Exposition, exhibit at centennial, 248; at Paris, 248, 256.

Fairbank, Samuel B., D. D., gifts of, 324.
Family feeling, 105, 118, 119, 120, 122; see *Homelike*.
Fast-days, 108, 151; special, 170; see *Day of Prayer*.
Fauth and Co., instruments by, 259.
Features of Mt. Holyoke Seminary, 36, 37, 54, 55, 327.
Felt, Rev. Joseph B., 39, 63.
Ferguson, Abbie P., 261, 323, 340.
Fever in 1840, 153.
Finances; receipts for building and furnishing, 91, 142, 143, 226, 238, 269; see *Gifts*; for improvements, obtained by students and teachers, 192, 222-224, 247, 258, 267, 268; by trustees, 193; grant from the state, 225; current expenses how met, 72, 73, 225, 281; see *Tuition*; present condition of, 279-281; funds needed, 281, 326.
Fire-escape, 248.
Fisher, James, 332, 333.

INDEX.

Fisher, Jeanette, 339.
Fisher, Samuel W., D. D., 189, 360.
Fiske, Catharine, 8.
Fiske, Fidelia, 205, 209, 219, 272, 320, 352; goes to Persia, 164-166; box for, 173; Dr. Perkins's estimate of, 164, 176; work in Persia, 328-330; letters to, 172, 174, 176; from letters of, 172, 173, 215; quoted, 19, 20, 90, 154, 162, 213, 217; returns to U. S., 202; impressions of the sem., 203, 204; invited to remain, 207; stimulates interest in missions, 204, 205, 210; labors for souls, 211, 212, 214, 215, 218; death, 216, Life of, vi.; reading of it in England, 344.
Fiske, Rev. Pliny, 23, 164.
Foote, Nancy A., Mrs. Webb, 174, 320, 352.
Foster, Mrs. Mary A. (Hurd), 253, 324, 359.
Foster, Sarah J., Mrs. Rhea, 320, 330, 331.
French, Helen M., principal, 221, 223, 225, 350, 352; administration of, 221-234; see *Gulliver*.
French, see *Languages*.
Fritcher, Ann Eliza, 209-211, 214, 216, 231, 320, 338, 352.
Funds, permanent, 279-281.

Georgia College, 256.
German; see *Languages*.
Gifts to the seminary; see *Alumnæ*; *Donors*; first $1,000, 40, 60; from friends in Boston, 68; Wheaton Seminary, 71, 85; agency of Dr. Packard, 51; of Mr. Hawks, 53, 56; records of, 58, 326, quoted, 59; obtained by students and teachers, see *Finances*.
Gifts of the sem. to missions, 24, 174.
Gilson, H. Juliette, 320, 323, 342.
Giving, inculcated, 161; see *Benevolence*.
"Good of the whole," 111, 121.
Goodnow, Hon. E. A., gifts of, 261, 342.
Goodnow Park, 261, 270, 271, 280.
Goodyear, Prof. Wm. Henry, M. A., 255.
Gordon, Alice W., Mrs. Gulick, 231, 320, 344, 352.
Gough, John B. 148.
Graham, Mrs. Harris, see *Bray*.
Grant, Mrs. Judith S., 328.
Grant, Z. P., at Byfield, 28; Derry, 30; Ipswich, 30-33; invited to Franklin county, 31; to join Miss Lyon at So. Hadley, 67; objections to of Holyoke plan, 43, 52; aids needy students, 43; letter of, 67, 68; see *Banister*.

Grave of Miss Lyon, 181, 272.
Gray, Prof. Asa, 248.
Greek, see *Languages*.
Greene, John M., D. D., quoted, 81, 226, 233; acknowledgment to, vi; pastor, 361.
Greenhouse, 268, 324.
Gridley, Eber, bequest of, 280.
Grounds of the seminary; original lot, 84, 65; additions, 224, 259, 261, 268, 269; see *Observatory*; *Goodnow Park*; *Botanical garden*; *Dwight place*; *Improvements*.
Gulick, Luther H., M D., 211.
Gulick, Rev. William H., 231.
Gulliver, Mrs. Lemuel; see *French*; 236, 278, 349.
Guyot, Prof. A. H., 238.
Gymnasium, 195, 196, 222, 224.

Habits cured, 125, 126; formed, 292, 293, 294, 295, 300, 305.
Hadley, 79; History of, quoted, 4, 81; first settlers, 14; street, 83.
Hampshire county, 14; "the banner county," 80.
Hampshire Gazette, quoted, 4.
Harding, Rev. Charles, 231; Mrs., 319; see *Ballantine*.
Harris, Samuel, D. D., LL.D., 181, 348, 360.
Harris, William T., LL.D., 269.
Hatfield, 4; see *Schools*.
Harvard; see *College*; *Harvard Register*, quoted, 85.
Haven, Mrs. Joseph, 43.
Hawes, Joel, D. D., addresses of, 100, 173, 200, 359, 360; quoted, 142.
Hawks, Rev. Roswell, aid of, 51-55, 175, 185, 188; agency of, 53, 56; trustee, 61, 348; steward, 191, 359; at Painesville. 54, 335; death, 232, 233; Mrs. Hawks, 53.
Hayes, Joel, 84, 89.
Hazeltine, Phebe, bequest of, 279.
Hazen, Allen, 211, Mrs., 211; see *Chapin*.
Hazen, Sophia D., 185; asso. prin., 186, 187, 350; 321, 352; see *Stoddard*.
Health, cared for, 104, 119, 124, 125.
Health of women, 103, 125.
Heard, George W., 39, 63, 65.
Hebrew, time for, desired, 149.
Herrick, John R., D. D., pastor and trustee, 245, 246, 349, 361.
Higginson, T. W., quoted, 4, 141.
Historical sketch of the sem., 248, 256.
Hitchcock, Prof. Charles H., Ph. D., 228, 250.

Hitchcock, Edward, D. D., LL.D., 80, 200, 204; aid of, 17, 32, 39, 48, 52; gift of, 194; trustee, 66, 348; lectures, 147, 197; addresses, 359, 360; quoted, 19, 22, 98, 124, 150; death, 217.
Hitchcock, Mrs. Edward, 17, 52; design of, for diploma, 101; death, 217.
Hitchcock, Edward, M. D., 229.
Hitchcock, Hon. Reuben, 337.
Homelike, the seminary, 230, 295, 314, 315; see *Family-feeling*.
Honor, sense of, cultivated, 112.
Hooker, Henry B., D.D., 208.
Hooker, Henrietta E., 273, 353.
Hopkins, Mark, D.D., LL.D., quoted, 9, 140; 359.
Hopkins, Catharine, asso prin., 191, 206, 208-211, 350; in charge of the sem., 218; illness and death, 218; 219, 353.
Houghton, Marilla, Mrs. Gallup, 335.
Howland, Gen. Asa, 39.
Huguenot Seminary, sketch of, 340.
Humphrey, Heman, D. D., 32, 66, 181, 360.
Hyde, Ira, 191, 359.

Imprint on alumnæ, 296, 304, 307-309.
Improvements in seminary grounds, 84, 119, 268, 274; shade trees and bridges, 191, 262; interior, 192, 193, 226, 247, 268, 269; see *Steam-heating; Water*.
Incorporation of Adams, Ipswich, and Abbot Academies, 7; Mt. Holyoke Seminary, 11, 61; Wheaton Seminary, 71; Harvard College, 11; Wellesley college, 239.
Indebtedness of students to the seminary, 114, 115, 144, 156; gratefully acknowledged, 289-316.
Indians, N. A., missionary teachers among, 24; see list, 319-322.
Ipswich Seminary, under Miss Grant, 30-37, 69; Mr. and Mrs. Cowles, 158; results of its training, 33, 41; efforts for endowment of, 31; effect of failure, 34; in charge of Miss Lyon, 36, 37, 69; relations to Mt. Holyoke Seminary, 54, 56, 69.

Jessup, Emily, associate principal, 189, 191, 350; 303, 313, 333, 334, 353; quoted, 115, 293, 304.
Johnson, Harriet, Mrs. Loughridge, 186, 321, 353.
Journal, the sem., begun, 172; quoted, 109, 177, 181, 184, 191-193, 230-232, 247, 248, 264, 267, 271-276, 277.
Judd, Sylvester, quoted, 81.
Judson, Ann Hasseltine, 28.

Kansas-Nebraska bill, 188.
Kelsey, Adaline D. H., M. D., 265, 321, 357.
Kendall, Henry, gift of, 280.
Kimball, Prof. Alonzo S., M. A., 270.
King, Jonas, D.D., 23.
Kingman, Abner, gift of, 196; trustee, 226, 348; death, 262.
Kirk, Edward N., D.D., meets Miss Lyon, 201, 167; benefactions, 193, 194, 224; trustee, 202, 348; labors for souls, 202, 212, 213, 264; addresses, 199, 202, 360; desires for the seminary, 240, 316; death, 241.
Knapp, Mrs. Alzina M., 339.

Laboratory work, 228, 238, 271.
Lake Erie Seminary, 190, 214, 235; sketch of, 335-338.
Languages taught, 148, 236, 267, 282; French, 148, 236, 244, 255, 282; German, 236, 244, 255, 283; Greek, 149, 237, 244, 282.
Latin, provided for, 146; required, 148; standard advanced, 192, 196; see *Course of study*.
Laurie, Thomas, D.D., 180, 181, 200, 323, 361; quoted, 25, 191; acknowledgment to, vi.
Lawrence, Ithiel, 278, 359; Mrs., 279.
Lecture courses, 147, 197, 228, 237, 249, 250, 255, 269, 270.
Lexington, normal school at, 9.
Library, 193, 226, 270; fund for, 270, 280.
Life of Mary Lyon, publishers of, v.; reading of in England, 344; in So. Africa, 340; translated in Africa, 341; in Japan, 346.
Long, Hon. John D., 260.
Lord, John, LL.D., 269.
Love, Wm. De Loss, D. D., 277, 349, 361.
Lyman, George S., succeeds "Cornelius," 275.
Lyman, Hannah, quoted, 24.
Lyon, Aaron 13, 15, 19.
Lyon, Electa, 18.
Lyon, Jemima (Shepard), 13, 15, 19, 20, 156.
Lyon, Lucy, Mrs. Lord, 171, 174, 175, 177, 321, 353, 358.
Lyon, Mary, sketch of life of, 12-26: ancestry, 13-15: birth-place, 12: heritage, 15: at school, 16; in Ashfield, 16, 17; Amherst, 16; Byfield, 17, 21: teaching, 1814-1834, 18: associated with Miss Grant, 18, 30-37, 69: not satisfied, 33: failure in efforts for Ipswich Sem., 32: plans for a permanent sem., 36: feelings in leaving

Ipswich, 37, 38: her aim, 22, 37, 66, 103; its breadth, 35, 47, 115, 149: raises $1,000, 40: plans opposed, 41-47, 146, 142; misunderstood, 42-47: sensitive to criticism, 41, 56: confident of success, 39, 43, 47, 56, 90, 115: gratitude to early friends, 63, 64, 143: at Wheaton Seminary, 70, 74, 75: toil in founding Mt. Hol. Sem., 39-90: theory of education and educational work, 44, 103, 105, 111, 116, 127, 128, 150; based on Christian benevolence, 35, 36, 62, 66, 91, 111, 112, 113; theory tested, 91-102: methods adopted, 103-116; illustrated, 117-137: enthusiasm in teaching, 22, 70, 100, 123, 124: religious character, 19-26, 117; conversion, 20; faith, 19-20; in prayer, 20, 47, 50, 58, 62, 64, 93, 151; in intercession, 38, 109, 142, 166-168: interest in missions, 23, 160, 170, 177; its breadth, 163, 311; "Missionary Offering," 170: sense of responsibility, 107, 114, 160: power over others, 88, 110, 120, 121, 126, 129, 136, 142; leads pupils to Christ, 22; see *Conversion; Revivals*; to mission work, 23, 24, 33; see *Missionary history*: compared with Dr. Arnold, 25: growth of the sem. during her life, 138-182: illness and death, 179-181: see *Life of Mary Lyon*.

Mann, Rev. Cyrus, 48.
Martyn, Henry, quoted, 150.
Mather, Cotton, quoted, 1.
Mather, Richard H., D. D., 255.
Maynard sisters, gift of, 58.
McCabe, Miss, 333.
McKeen, Catharine, 194, 196, 296, 353.
Mead, Hiram, D. D., pastor and trustee, 198, 199, 206, 208, 348, 361.
Mears, Leverett, Ph. D., 270.
Meetings, see *Religious instruction*.
Melvin, Helen E., 278, 321, 354.
Melvin, Sarah H., 326, 354.
Memorandum Society, 255, 318, 326, vi.
Merriam, Homer, gift of, 279.
Methods adopted, 103-116; illustrated, 117-137, 273-276, 291, 297; self-help, 103; sanitary, 104, 124; plan of instruction, 104, 105; see *Course of study*; use of Bible, 105-107, 301; prayer, 107-109; work specific, 109; personal influence, 110, 117, 121, 126, 129; teachers co-operate, 110; self-government, 111; cultivation of conscience, 111, 112; honor, 112; benevolence, 113; responsibility, 114.

Michigan Seminary, sketch of, 339.
Midland Seminary, 343.
Microscopical outfit, 270.
Mills, Cyrus T., D. D., 338, Mrs. Susan, 338; see *Tolman*.
Mills Seminary and College, sketch of, 338-339.
Missionaries from the Buckland school, 33; Ipswich Sem., 24; Mt. Hol. Sem., see *Alumnæ*; under foreign boards, list of, 319-322; from Holyoke faculty, 164, 174, 177, 209, 216, 231, 265, 278; summary, 24, 318, 323.
Missionary history, 159-166, 170-177, 204-211, 215, 216, 220, 231, 265, 277, 278; instruction on missions, 160, 171, 177, 201, 232; meetings, 160, 171, 201, 232, 277; testimony of alumnæ, 291, 310-312.
"Missionary offering," 170, 210; quoted, 13, 114.
Missionary reunion, 204.
Montague, Mrs. Obed, quoted, 77, 101.
Montague, Elliot and Newton, 89.
Moody, D. L., 230, 277, 346.
Moody, Pliny, 80, 83.
Moore, Abigail, 148, 354; associate principal, 159, 350; goes to India, 174, 185, 321; see *Burgess*.
Moral training paramount, 103, 115, 116, 128, 317.
Morning exercises, see *Religious instruction*.
Morrow, Rev. Horatio, 265, Mrs. O. J. (Emerson), M. D., 265.
Mortality of alumnæ, 229.
Mount Holyoke, 49, 83, 271.
Mount Holyoke Seminary, a pioneer, 1, 18, 36, 37, 66, 72, 189, 190, 240, 327: anticipated, 27-38: plan for founding, 36, 39, 44-46; foundation principle, 19, 35, 36, 52-54, 72, 91: object of, 35, 54, 55, 66, 72, 103: for whom designed, 33, 34, 45, 54, 56, 103: preparations, 39-60; places for, proposed, 48: located, 46, 48: obstacles, 41-46, 138, 140, 142: ridiculed, 41, 46, 49, 53: misunderstood, 44, 55: named, 48, 49, 149: early friends, 39, 46, 49, 51, 57, 58, 62-64: subscriptions, 46, 59, 60, 68, 71; see *Gifts; Donors*: charter, 11, 61: erection of first building, 61-77; see *Buildings*: consecrated, 58, 76, 90, 93, 145, 162: opened, 78-90: literary standard, 54, 56; see *Course of study*: permanent, 43, 54, 73, 116, 149, 184, 316: progressive, 56, 105, 115, 145, 149, 228, 236, 249, 267, 271, 315: theory

tested, 91-102; see *Domestic department*: methods adopted, 103-116; illustrated, 117-137; ideas embodied v., 327; essential features, 37, 54, 55, 327; growth under Miss Lyon, 139-183; Miss Chapin, 184-220; Miss French, 221-234; Miss Ward, 235-267; Miss Blanchard, 268-283; unsectarian, 110, 306; statistics of attendance, 284-287; results, testimony of alumnæ, 288-317; work of alumnæ, 318-326; similar institutions, 327-347; trustees, 348; teachers, 349-358; see *Donors; Finances*.
Mount Tom, 83, 271.
Munsell, Esther E., Mrs. Thompson, 205, 321, 355.
Music, 146, 149, 197, 255, 276; private instruction, 268, 283.
Mutual influence, 109, 122, 123, 306.
Murray, Rev. Andrew, 340, 342, 344, Mrs., 340.
Murray, Rev. Charles, 343.
Murray, Helen, 343.
Murray, Rev. William, 343.

Natural History Society, 228.
Neethling, Rev. J., 342.
Newcomb, Simon, LL. D., 261.
Newell, Harriet Atwood, 28.
Newton, Martha C., 323, 343.
Norcross Roseltha A., 220, 321.
Normal school, first in U. S., 9, 160.
Notch, the, 83.
Number of applicants, 89, 94, 138, 186, 192; of students, 88, 94, 138, 141, 155, 173, 186, 192, 240, 268, 269; see *Statistics*.
Nutting, Mary O., 228, 248, vi.

Oberlin Collegiate Institute, 9, 278.
Observatory, the, 193; memorial, 237, 239, 258; described, 259; dedicated, 260; 273.
Optional courses, 271, 282; studies, 271.
Oroomiah, sem. at, 176, 247; sketch of, 328-331; see *Fiske, Fidelia*.
Outdoor exercise, 272, 247.
Oxford College. O., 333.

Packard, Theophilus, D. D., 30, 63; aid of, 39, 42, 50, 51.
Painting, 274, 283.
"Paradise," 82, 84.
Park, E. A., D. D., LL. D., 209, 360.
Parmelee, Olive L., Mrs. Andrus, 231, 321, 354.
Parsons, Rev. H. M., 231.

Parsons, Rev. Levi, 23.
Pavilion, the, 262, 270, 272.
Peabody, Andrew P., D. D., LL. D., quoted, 87.
Peabody, Helen, 332-334, 354.
Pease, Claudius B., 220; Mrs., see *Chapin, Mary W.*; vi. 255, 318, 319.
Penikese, 238.
Penney, Joseph, D. D., 39, 55.
Perkins, Justin, D. D., visit of, 163 328; 171, 177; quoted, 164, 176; Mrs., 163.
Philbrick, J. D., LL. D., 256.
Phillips, David E., 279, 359.
Phillipston, letter from, 74.
Piano practice, 148, 283.
Piety and scholarship, 150.
Political science, 147, 271.
Pond, Clara C., Mrs. Williams, 211, 216, 322, 354.
Porter, Andrew W., aid of, 61-64, 76, 87, 144, 179; gifts of, 63, 194, 196; trustee, 62, 225, 348; last visit, 252; death, 253; children of, 63.
Porter, Mrs. Hannah, 62, 87, 196, 232; gift of, 196; quoted, 62, 74, 75, 179.
Porter, Mrs. Mary (Stafford), 253, 254.
Porter, Catharine A., Mrs. Pitkin, 174, 354.
Porter, Lewis H , 279, 359.
Portraits in seminary hall, 272.
Postal rates in 1837, 57, 74.
Post-graduate courses, 267, 271, 282.
Post-offices in 1790, 3.
Pratt, Capt., Carlisle, Pa., 292.
Prayer, seminary born of, 55, 58, 107; taught, 107, 108; time for, secured, 107, 276; testimony of alumnæ, 289, 291, 399; days of, 108; see *Fast-days*; specific, 151, 161; for alumnæ, 109; at Sabbath twilight, 154, 157; meetings for, see *Religious instructions*; for soldiers, 198, 207; answered, 50, 62, 63, 64, 65, 94, 102; see *Revivals; Conversions; Lyon, Mary*.
Principal, ability required in, 184, 186, 187.
Prospect Hill, 80, 82, 261; see *Goodnow Park*; in South Africa, 341.
Putnam, Mrs. Elizabeth S. (Hawks), 233.

Quarter centennial, 140, 142, 199.
Quincy's "Municipal History of Boston," quoted, 4.

"Recess meetings," 108, 157, 160, 276; described, 158.
Reed, Susan, Mrs. Howland, 174, 322, 354.
Reed, V. D., D. D., 214, 360.

Religious controversy, in Connecticut, 13; in Hampshire county, 14.
Religious history of the seminary, 150-183, 204-216, 230-232, 263-265, 276.
Religious instruction, 105-107, 136, 231, 232, 234; feelings of Miss Lyon concerning, 107; of others, 175, 212; morning exercise, 105, 130, 132, 188, 221, 272, 290; meetings, daily, 108; see *Recess;* Sabbath, 106, 108; Thursday, 106, 108, 276, 277; inquiry, 106; testimony of alumnæ, 132, 290, 295, 305-310; see *Bible.*
Requirements for admission, 145, 281.
Results of seminary training testified, 289-316; appreciated gradually, 297.
Revivals, at Buckland, Derry and Ipswich, 22; first at Holyoke Seminary, 152; in 1843, 166-170; in 1864, 212-214; in 1879, 231; in 1884, 276; character of, 150, 151; means used, 151, 153, 155, 166-169, 201; see *Conversions.*
Revolution, effect of on education, 4.
Reynolds, Mary E., 210, 211.
Rhea, Mrs S. J. (Foster), 329, 330, 331.
Rice, Col. Austin, 262, 348.
Rice, Mary S., 177, 322, 329, 330, 331.
Rice, Prof. William N , 269.
Riggs, Mrs. Stephen, Mary Ann Longley, 33.
Ross, John, 331.
Rowing provided for, 247.
Rules, 111, 119, 121, 125, 276.

Sabbath, the, 108, 109; letter writing on, 108, 127.
Safford, Hon. Daniel, aid of, 50, 63, 64, 76, 77, 87, 144, 194, 201, 202; gifts of, 50, 68, 83, 89, 193-195; trustee, 66, 348; death, 194.
Safford, Mrs., 49, 67, 76, 87, 201, 202; death, 211.
Sawyer, Hon Edmund H., 251, 253, 256, 348; Mrs., gift of, 251.
School laws of 1642 and 1647, 2; of 1789, 5.
Schools, early provision for, 1; grammar, object of, 2; see *Colleges;* public, not attended by girls, 3; girls admitted, 4, 5; branches taught, 2, 3, 5, 6; for girls, expensive, 9, 10, 33, 34; why not permanent, 9, 29, 31; for Catholic girls, 10, 11, 72; see *Education.*
"Sections," 108, 109, 276.
Sedgwick, William T., Ph D , 271.
Seelye, J. H., D. D., LL. D., 246, 247, 260, 348, 360.

Self-denial, spirit of, inculcated, 23, 127, 162, 163, 291, 299, 305, 306, 312, 313.
Self-government, 111, 112.
Self-help, 15, 16, 36, 44, 103.
Self-reporting, 112, 291, 304.
Sessions, Lydia A., Mrs. Woodworth, 336, 355.
Seymour, Harriet, 229.
Shattuck, Lydia W., 225, 227, 238, 247, 324, 355; effort for elevator, 258; reminiscence of, 313; quoted, 323.
Shepard, Isaac, 15; Mrs. Jemima (Smith), 15.
Similar institutions, sketches of, 327-347: in Persia, 176, 214, 328-331; see *Fiske, Fidelia:* Tahlequah, 186, 331: Ohio, Oxford, 189, 214; see *Western Seminary, The;* Painesville, 190, 214; see *Lake Erie Seminary:* Marsovan, and Bitlis, Turkey, 338, 339: California, 338; see *Mills Seminary and College:* Michigan, 339: South Africa, 340-344: Spain, 344-346: Japan, 346: Minnesota, 347.
Site, selected, 40; changed, 64, 65.
Skepticism precluded, 300, 301.
Smith, Abigail Tenney, 33.
Smith, Annie, 343.
Smith, Byron, 89, 199.
Smith, Chileab, in South Hadley and Ashfield, 14; character of, 14; descendants, 15; longevity, 15.
Smith, Rev. Henry, Lieut. Samuel, and other ancestors of Mary Lyon, 13.
Smith, A. Dennison, 14.
Smith, Edward, 279.
Smith, Ellen A., 323, 343.
Smith, E. T., Esq., 199.
Smith, G. Morgan, 89.
Smith, Mrs. Moses, Emily A. White, 326.
Smith, Rev. Preserved, 334.
Smith, Sophie A., 278, 355; see *Burt.*
Smith, see *College.*
Snell, Ebenezer S., LL. D., 147, 197.
Soldiers, work and gifts for, 198; prayer-meeting, 198, 207.
South Hadley, in 1820, 81; in 1837, 82; early settlers of, 14, 81; fossil foot marks, 80, 83; scenic beauty of, 65, 82, 140; ladies of, help Miss Lyon, 77; kindness of citizens, 89; hospitality of, 88 199; village, fire in, 244, 248.
South Hadley Sabbath-school, semi-centennial of, 81.
Southworth, Hon. Edward, 233.
Spanish Holyoke Seminary, sketch of, 344-6.
Spaulding, Mrs. Julia (Brooks), 33.

Specific labor, 109.
Spofford, Sophia, 188, 189, 234, 290, 350, 355.
Spooner, Mary Ella, 278, 355.
Stafford, Mrs. Mary, 254; see *Porter*.
Start, Sarah A., 302, 303, 312, 355.
Statistics of attendance, 284-287; of Memorandum Society, 318.
Steam-heating, 222-226, 252.
Stevens, Mary M., 148, 355, 359.
Stewards, list of, 360.
Stoddard, Charles, 49.
Stoddard, Rev David T., 186.
Stoddard, Mrs., see *Hazen, Sophia D.*, 202, 218; acting prin., 219; 350, 355.
Stoddard, Sarah, 202, 203.
Stony Brook, 80, 82, 84, 191, 257; in South Africa, 341.
Storrs, Otis, 4.
Storrs, Richard S., D. D., LL. D., 200, 360.
Stuart, George H., 198.
Swift, Rev. E. Y., pastor and trustee, 181, 193, 200, 206, 348, 361.

Tahlequah, Ind. Ter., seminary at, 186, 331.
Teachers, counsels to, 127; wages of, 4, 5; women, legally recognized, 5; ratio of women to men, 10; at Mt. Holyoke, 69, 91, 110; meetings of, 108, 109; go to foreign fields, 164, 174, 177, 210, 216, 231, 265, 278.
Temperance Union, 277.
Tenney, Rev. Daniel, 200, 332; Mrs., 332.
Testimony of alumnæ, 288-316 ; see *Bible ; Missionary meetings; Religious instruction:* of others, Dr. Fisher, 189; Miss Fiske, 203, 204; Mrs. Banister, 243; Dr. Kirk, 316.
Thanksgiving evening, 192, 198.
Thayer, Sarah A., 323, 343, 359.
Theory, Miss Lyon's, of piety and scholarship, 150; of education, 44, 103, 105, 111, 116, 127, 128, 150; based on benevolence, see *Lyon, Mary*.
Theory tested, 91-102, 150, 288-347.
"Thermopylæ," 83.
Thompson, Rev. Amherst L., 205; Mrs. Esther E. (Munsell), 205, 321, 355.
Thompson, Prof Charles O., 237.
Thoroughness required, 104, 123, 290, 295, 300, 361.
Thurston, Persis, Mrs. Taylor, 310, 311, 322.
Tichenor, Gabriel, 332, 333.
"Titan's Piazza" and "Titan's Pier," 83.

Todd, John, D. D., 39, 42, 49, 55, 61, 65, 348.
Tolman, Julia M., Mrs. Lucius A., asso. prin., 190, 280, 350; 335, 356.
Tolman, Susan M., 177, 322, 356 ; see *Mills*.
Traveling facilities in 1837, 78.
Treat, Rev. Selah B., 209.
Troy Seminary, 8.
Trustees, co-operation of, 64, 184; list of, 348; women elected, 278.
Tuition, why put low, 34, 40; first year, 72, 73; next sixteen years, 100; increased, 197, 226; table, 284; see *Finances*.
Tyler, Rev. William, 39, 61, 64.
Tyler, William S., D. D., LL. D., trustee, 243, 348, 349; addresses by, 251, 360; quoted, 80, 240, 243, 254, 266; acknowledgment to, vi.

Underwood, Rev. Rufus S., 277.
Unsectarian, 110, 306.
Unselfish spirit, training in, 111, 114, 128, 162, 163; shown, 69, 99, 186; see *Benevolence; Self-denial*.
United States Christian Commission, 198.

Vassar College, 256.
Venus, transit observed, 260, 261.

Wages of teachers, 4, 5; views of Miss Beecher and Miss Grant, 43; of Miss Lyon, 35, 36, 43-46, 91.
Ward, Julia E., asso. prin., 201, 234; prin., 235, 350; 242, 252, 258, 261, 262; resigns, 266; 267, 268, 269, 356; administration of, 235-267.
Warfield, Mary E., 220.
Warren, Rev. I. P., 200.
War-time, seminary in, 197, 207, 211; see *Soldiers*.
Washburn, Fannie E., 231, 322, 356.
Washburn, Ruth, see *Bancroft*.
Water in each story, 191; tower, 196; from artesian well, 257; water power, 269.
Watson, Laura C., 347.
W. C. T. U., auxiliary, 277.
Wellesley, see *Colleges*.
Wells, Annie M., 265, 323, 356.
Western Seminary, The, 189, 214; sketch of, 332-335.
Wheaton, Judge, 69.
Wheaton Seminary, founded, 69; aid of Miss Lyon, 70, 71; incorporated, 71; gifts of to Mt. Holyoke, 71, 85.

Wheeler, Rev. Melancthon G., 46.
White, Mary A., 33.
White, Rev. Morris, 42.
Whitman, Mary C., asso. prin., 159, 175, 350; prin , 185-187, 349, 356; see *Eddy*; quoted, 158, 168, 178, 201; letter to, 173; father of, 187.
Wilder, Eliza A., Mrs. Holmes, 197, 357.
Willard, Mrs. Emma, 8.
Williston. A. Lyman, M. A., trustee, 348; treasurer. 349; 246, 253; gifts of, 238, 252, 259, 261, 342.
Williston, Mrs. A. Lyman, gift of, 251; trustee, 278, 349.
Williston, John Payson, 260.
Williston Hall, 237-239, 246, 250, 252, 272-274.

Williston, Hon. Samuel, 40, 61; gifts of, 80, 196.
Willoughby Seminary, 190, 335.
Wilson, Prof. Edmund Beecher, 271.
Wilson, James D , D. D., quoted, 288.
Winslow, Mrs. Myron, Mary Billings, 33.
Woman's Boards of Missions, 23, 177, 312.
Woodbridge, William, first advanced school for girls, 7.
Woodbridge school, South Hadley, 82.
Wright, Lucy, Mrs Mitchell, 322, 350.
Wright, Abby, school of, 82.

Yohanan, Mar, 328.
Young, Charles A., Ph. D., LL. D., 228, 259, 260, 349.

www.ingramcontent.com/pod-product-compliance
Lightning Source LLC
Chambersburg PA
CBHW030551300426
44111CB00009B/942